Painted by F. Winterhalter.

H. M. G. Majesty, Victoria, Queen of Great Britain,

IN THE ROBES OF THE MOST NOBLE ORDER OF THE GARTER.

DAGUERREOTYPED BY THOMPSON.

From the Portrait in possession of Geo. P. Burnham;

PRESENTED TO HIM BY HER MAJESTY, THE QUEEN, IN 1853.

[See Letter, page 130.]

THE HISTORY OF THE HEN FEVER.

A Humorous Record.

BY
GEO. P. BURNHAM.

In one Volume.—Illustrated

BOSTON:
JAMES FRENCH AND COMPANY.
NEW YORK: J. C. DERBY.
PHILADELPHIA: T. B. PETERSON.

STEREOTYPED BY
HOBART & ROBBINS,
New England Type and Stereotype Foundery,
BOSTON.

GEO. C. RAND, PRINTER, 3 CORNHILL.

TO THE

Amateurs, Fanciers, and Breeders

OF

POULTRY,

THE SUCCESSFUL AND UNFORTUNATE DEALERS,

THROUGHOUT THE UNITED STATES;

AND

THE VICTIMS OF MISPLACED CONFIDENCE IN

THE HEN TRADE, GENERALLY,

I DEDICATE

This Volume.

PREFACE.

In preparing the following pages, I have had the opportunity to inform myself pretty accurately regarding the ramifications of the subject upon which I have written herein; and I have endeavored to avoid setting down "aught in malice" in this "*History of the* Hen Fever" in the United States.

I have followed this extraordinary *mania* from its incipient stages to its final death, or its *cure*, as the reader may elect to term its conclusion. The first symptoms of the fever were exhibited in my own house at Roxbury, Mass., early in the summer of 1849. From that time down to the opening of 1855 (or rather to the winter of 1854), I have been rather intimately connected with the movement, if common report speaks correctly; and I believe I have seen as much of the tricks of this trade as one usually meets with in the course of a single natural life.

Now that the most serious effects of this (for six years) alarming epidemic have passed away from among us, and when "the people" who have been called upon to pay the cost of its support, and for the burial of its victims, can look back upon the scenes

1*

that have in that period transpired with a disposition cooled by experience, I have thought that a volume like this might prove acceptable to the hundreds and thousands of those who once "took an interest in the hen trade," — who *may* have been mortally wounded, or haply who have escaped with only a broken wing; and who will not object to learn how the thing has been done, and "who threw the bricks"!

If my readers shall be edified and amused with the perusal of this work as much as I have been in recalling these past scenes while writing it, I am content that I have not thrown the powder away. I have written it in perfect good-nature, with the design to gratify its readers, and to offend no man living.

And trusting that *all* will be pleased who may devote an hour to its pages, while at the same time I indulge the hope that *none* will feel aggrieved by its tone, or its text, I submit this book to the public.

Respectfully,

GEO. P. BURNHAM.

RUSSET HOUSE, *Melrose*, 1855.

CONTENTS.

THE

HISTORY OF THE HEN FEVER.

CHAPTER I.

PREMONITORY SYMPTOMS OF THE DISEASE.

I WAS sitting, one afternoon, in the summer of 1849, in my little parlor, at Roxbury, conversing with a friend, leisurely, when he suddenly rose, and passing to the rear window of the room, remarked to me, with considerable enthusiasm,

"What a splendid lot of fowls you have, B——! Upon my word, those are very fine indeed,—do you know it?"

I had then been breeding poultry (for my own amusement) many years; and the specimens I chanced at that time to possess were rather even in color, and of good size; but were only such as any one might have had — bred from the common stock of the country — who had taken the same pains that I did with mine.

There were perhaps a dozen birds, at the time, in the

rear yard, and my friend (*then*, but who subsequently passed to a competitor, and eventually turned into a sharp but harmless enemy) was greatly delighted with them, as I saw from his enthusiastic conversation, and his laudation of their merits.

I am not very fast, perhaps, to appreciate the drift of a man's motives in casual conversation,— and then, again, it may be that I am "not so slow" to comprehend certain matters as I might be! At all events, I have sometimes flattered myself that, on occasions like this, I can "see as far into a millstone as can he who picks it;" and so I listened to my friend, heard all he had to say, and made up my mind accordingly, before he left me.

"I tell you, B——, those are handsome chickens," he insisted. "I 've got a fine lot, myself. You keep but one variety, I notice. I 've got 'em *all*."

"All what?" I inquired.

"O, all kinds — all kinds. The Chinese, and the Malays, and the Gypsies, and the Chittaprats, and the Wang Hongs, and the Yankee Games, and Bengallers, and Cropple-crowns, and Creepers, and Top-knots, and Gold Pheasants, and Buff Dorkings, and English Games, and Black Spanish and Bantams,— and I 've several *new breeds* too, I have made myself, by crossing and mixing, *in the last year*, which beat the world for beauty and size, and excellence of quality."

"Indeed!" I exclaimed. "So you have made several new *breeds* during *one* year's crossing, eh? That *is* remarkable, doctor, certainly. I have never been able yet to accomplish so extraordinary a feat, myself," I added.

"Well, *I* have," said the doctor,—and probably, as he was a practising physician of several years' experience, he knew how this reversion of nature's law could be accomplished. I did n't.

"Yes," he continued; "I have made a breed I call the 'Plymouth Rocks,'—superb birds, and great layers. The —a—'Yankee Games,'—regular knock-'em-downs,— rather fight than eat, any time; and never flinch from the puncture of steel. Indeed, *so* plucky are these fowls, that I think they rather *like* to be cut up than otherwise,— alive, I mean. Then, I 've another breed I 've made—the 'Bengal Mountain Games.' These *are* smashers—never yield, and are magnificent in color. Then I have the '*Fawn-colored Dorkings*,' too; and several other fancy breeds, that I 've fixed up; and fancy poultry is going to sell well in the next three years, you may be sure. Come and see my stock, B——, won't you? And I 'll send you anything you want from it, with pleasure."

I was then the editor of a weekly paper in Boston. I accepted my friend's kind invitation, and travelled forty miles and back to examine his poultry. It looked well— *very* well; the arrangement of his houses, &c., was good, and I was gratified with the show of stock, and with his politeness. But he was an enthusiast; and I saw this at the outset. And though I heard all he had to say, I could not, for the life of me, comprehend how it was that he could have decided upon the astounding merits of all these different *breeds* of fowls in so short a space of time — to wit, by the crossings in a single year! But that was his affair, not mine. He was getting his fancy poultry ready for the market; and he repeated, "It will *sell*, by and by."

And I believe it did, too! The doctor was right in *this* particular.

He informed me that he intended to exhibit several specimens of his fowls, shortly, in Boston; and soon afterwards I met with an advertisement in one of the agricultural weeklies, signed by my friend the doctor, the substance of which was as follows:

NOTICE. — I will exhibit, at *Quincy Market*, Boston, in a few days, sample pairs of my fowls, of the following pure breeds; namely, Cochin-China, Yellow Shanghae, Black Spanish, Fawn-colored Dorkings, Plymouth Rocks, White Dorkings, Wild Indian, Malays, Golden Hamburgs, Black Polands, Games, &c. &c.; and I shall be happy to see the stock of other fanciers, at the above place, to compare notes, etc. etc.

The above was the substance of the "notice" referred to; and the doctor, coming to Boston shortly after, called upon me. I showed him the impropriety of this movement at once, and suggested that some spot other than Quincy Market should be chosen for the proposed exhibition,— in which I would join, provided an appropriate place should be selected.

After talking the matter over again, application was made to an agricultural warehouse in Ann-street, or Blackstone-street, I believe; the keepers of which saw the advantages that must accrue to themselves by such a show (which would necessarily draw together a great many strangers, out of whom they might subsequently make customers); but, at my suggestion, this very stupid plan was abandoned — even after the advertisements were circulated that such an exhibition would come off there.

Upon final consideration, it was determined that the first Exhibition of Fancy Poultry in the United States of America should take place in November, 1849, at the *Public Garden*, Boston.

CHAPTER II.

THE "COCHIN-CHINAS." BUBBLE NUMBER ONE.

A PUBLIC meeting was soon called at the legislative hall of the Statehouse, in Boston, which had the effect of drawing together a very goodly company of savans, honest farmers, amateurs, poulterers, doctors, lawyers, flats, fanciers and *humbugs* of one kind or another. *I* never attended one of the meetings; and only know, from subsequent public and private "reports," what occurred there.

On this *first* occasion, however, after a great deal of bosh and stuff, from the lips of old men and young men, who possessed not the slightest possible shadow of practical knowledge of the subject proposed to be discussed, it was finally resolved that the name for the (now defunct) association then and there formed, should be "*The New England Society for the Improvement of Domestic Poultry*" *!!!*

Now, the only objection I ever raised to this title was that it was not sufficiently *lengthy!* When applied to for my own views on the subject, *I* recommended that it should be called the "Mutual Admiration Society." But, though I was thought a great deal of by its members,— especially when the concern was short of funds,— in *this* case they thought my proposed title was altogether too applicable; and the original name, above quoted, was adhered to.

I was honored with the office of vice-president of the society, for Massachusetts; to which place I was reëlected annually, I believe, until the period of its death. For which honor I was not ungrateful, and in consideration of which, "as in duty bound, I have ever prayed" for the association's prosperity and weal.

The first name that was placed upon the list of subscribers to the constitution of this society was that of His Excellency Geo. N. Briggs, formerly Governor of this commonwealth. He was followed by a long list of "mourners," most of whom probably ascertained, within five years from the hour when they subscribed to this roll, that causing the cock's spur *to grow between his eyes* was not quite so easy a thing to accomplish as one "experienced poultry-breeder" at this meeting coolly asserted it to be! How many attempted this experiment (as well as numerous others there suggested as feasible), I am not advised. But I am inclined to think that those who did try it found it to be "all in their eye."

While these gentlemen were arranging the details of the new "society," and were deciding upon what the duties of the officers and committees should be, I quietly wrote out to England for information regarding the somewhat notorious "*Cochin-China*" fowl, then creating considerable stir among fanciers in Great Britain; and soon learned that I could procure them, in their purity, from a gentleman in Dublin, whose stock had been obtained, through Lord Heytsbury (then Lord Lieutenant of Ireland), direct from Queen Victoria's samples. I ordered six of them,— two cocks and four hens,—and in December, 1849, I received them through Adams & Co.'s Transatlantic Express.

At this period there was no telegraph established from Boston to Halifax, I believe. Some of the reporters for the daily city papers usually visited the steamers, upon their arrival here, to obtain their foreign files of exchanges; and here my birds were first seen by those gentlemen who have made or broken the prospects of more than one enterprise of far greater consequence than this "importation of fancy fowls" could seem to be.

But on the day succeeding the coming of those birds, several very handsome notices of the arrival of these august Chinamen appeared in the Boston papers, and a vast amount of credit was accorded to the "enterprising importer" of the outlandish brutes, that were described in almost celestial language!

After considerable trouble and swearing (custom-house swearing, I mean), the officers on board permitted my team to take the cage out of the steamer, and it was conveyed to my residence in Roxbury, where it arrived two hours after dark.

I had long been looking for the coming of these Celestial strangers, and the "fever," which I had originally taken in a very kindly way, had by this time affected me rather seriously. I imagined I had a fortune on board that steamer. I looked forward with excited ideas to beholding something that this part of the world had never yet seen, and which would surely astound "the people," when I could have the opportunity to show up my rare prize,— all the way from the yards or walks of royalty itself! I waited and watched, with anxious solicitude,— and, at last, the box arrived at my house. It was a curiously-built box — the fashion of it was unique, and substantial, and foreign in its exterior. I supposed, naturally, that its contents must be similar in character. That box contained my "Cochin-Chinas,"— bred from the Queen's stock,— about which, for many weeks, I had been so seriously disturbed.

I am *now* well satisfied that the "Cochin-China" variety of fowl is a gross fable. If such a breed exist, in reality, we have never had them in this country. Anything (and everything) has been *called* by this name among us, in the last five years; but the engraving on the following page, in my estimation (and I have been there!), is the nearest thing possible to a likeness of this *long* petted bird; and will be recognized, I think, by more than one victim, as an

2*

PORTRAIT OF THE "COCHIN-CHINA" FOWL!

accurate and faithful portrait of this lauded "magnificent" and "superb" bird!

I was anxious to examine my celestial friends at once. I caused the box to be taken into a shed, at the rear of the house, and I tore from its front a piece of canvas that concealed them from view, to behold a ——— well! *n'importe* — they were *Cochin-China* fowls!

But, since God made me, I never beheld six *such* birds before, or since! They resembled *giraffes* much more nearly than they did any other thing, carnivorous, omnivorous,— fish, flesh, or fowl. I let them out upon the floor, and one of the cocks seized lustily upon my India-rubber over-shoe, and would have swallowed it (and myself), for aught I know, had not a friend who stood by seized him, and absolutely choked him off!

This is truth, strange as it may seem; but I presume they had scarcely been fed at all upon their fortnight's voyage from Dublin, and I never saw any animals so miserably low in flesh, in my life, before. What with their long necks, and longer legs, and their wretchedly starved condition, I never wondered that the friendly reporters spoke of their appearance as being "extraordinary, and strikingly peculiar."

These were the original "Cochin-China" fowls of America. And they probably never had the first drop of Chinese blood in their veins, any more than had the man who bred them, and who knew this fact much better than I did — who knew it well enough.

I housed my "prize" forthwith, however, and provided them with everything for their convenience and comfort. The six fowls cost me ninety dollars. They *were* beauties, to be sure! When I informed a neighbor of their cost, he ventured upon the expressive rejoinder that I "was a bigger d——d fool than he had ever taken me for."

To which I responded nothing, for I rather agreed with him myself!

Nine months afterwards, however, I sold him a cock and three pullets, four months old, raised from those very fowls, for sixty-five dollars; and I did n't retort upon him even *then*, but took his money. The chickens I sold him were "dog-cheap," at that!

CHAPTER III.

THE FIRST FOWL-SHOW IN BOSTON.

NEVER in the history of modern "bubbles," probably, did *any* mania exceed in ridiculousness or ludicrousness, or in the number of its victims surpass this inexplicable humbug, the "hen fever."

Kings and queens and nobility, senators and governors, mayors and councilmen, ministers, doctors and lawyers, merchants and tradesmen, the aristocrat and the humble, farmers and mechanics, gentlemen and commoners, old men and young men, women and children, rich and poor, white, black and gray,—*everybody* was more or less seriously affected by this curious epidemic.

The press of the country, far and near, was alive with accounts of "extraordinary pullets," "enormous eggs" (laid on the tables of the editors), "astounding prices" obtained for individual specimens of rare poultry; and all sorts of people, of every trade and profession and calling in life, were on the *qui vive*, and joined in the hue-and-cry, regarding the suddenly and newly ascertained fact that hens laid eggs —— *sometimes;* or, that somebody's crower was heavier, larger, or higher on the legs (and consequently higher in value), than somebody else's crower. And the first exhibition of the society with the long name came off duly, at last, as agreed upon by the people, *and* myself.

"The *people*"*!* By this term is ordinarily meant the body-politic, the multitude, the citizens at large, the voters, the — the — a — the masses; the —— well, no matter! At the period of which I am now writing, the term signified the "hen-men." This covered the whole ground, at that time. Everybody was included, and thus nobody was left outside!

At this first show, the committee "flattered themselves" (and who ever heard of, or from, a committee that did n't do this!) that never, within the memory of the oldest inhabitant,— who, by the way, was then living, but has since departed to that bourn from which even defunct hen-men do not return,— never had such a display been witnessed; never had the feathered race before appeared in such pristine beauty; never had any such exhibition been seen or read of, since the world begun! And, to say truth, it was n't a very bad sight,— that same first hen-show in Boston.

Thousands upon thousands visited it, the newspapers appropriated column after column to its laudation, and all sorts of people flocked to the Public Garden to behold the "rare and curious and inexpressibly-beautiful samples" of poultry caged up there, every individual specimen of which had, up to that hour, been straggling and starving in the yards of "the people" about Boston (they and their progeny) for years and years before, unknown, unhonored and unsung.

Gilded complimentary cards, in beautifully-embossed envelopes, were duly forwarded by the "committee" to all "our first men," who came on foot or in carriages, with their lovely wives and pretty children, to see the extraordinary sight. The city fathers, the public functionaries, governors, senators, representatives, all responded to the invitation, and everybody was there.

The cocks crowed lustily, the hens cackled musically, the ducks quacked sweetly, the geese hissed beautifully, the chickens peeped delightfully, the gentlemen talked gravely, the ladies smiled beneficently, the children laughed joyfully, the uninitiated gaped marvellously, the crowd conversed wisely, the few knowing ones chuckled quietly, — everybody enjoyed the thing immensely,— and suddenly, prominent among the throng of admirers present, loomed up the stalwart form and noble head of Daniel Webster, who came, like the rest, to *see* what he had only "read of" for the six months previously.

The committee saw him, and they instantly lighted on him for a speech; but he declined.

"Only a few words!" prayed one of them.

"One word, *one* word!" insisted the chairman.

"I can't!" said Daniel.

But they were importunate and unyielding, that enthusiastic committee.

"Gentlemen!" said the honorable senator, at last, amid the din. "Ladies and gentlemen!" he continued, as a monster upon feathered stilts, at his elbow, shrieked out an unearthly crow, that drowned the sound of his voice instanter, — "Ladies and gentlemen, really — I — would — but the noise and confusion is so great, that I cannot be heard!" — and a roar followed this capital hit, that drowned, for the moment, at least, even the rattling, crashing, bellowing, squeaking *music* of the feathered bipeds around him.

The exhibition lasted three days. Unheard-of prices were asked, and readily paid, for all sorts of fowls; most of those sold being mongrels, however. As high as thirteen dollars was paid by one man (who soon afterwards became an inmate of a lunatic asylum) for a single pair of domestic fowls. It was monstrous, ridiculous, outrageous, exclaimed every one, when this fact — the absolute paying down of thirteen round dollars, then the price of two barrels of good wheat flour — was announced as having been squandered for a single pair of chickens.

I *sold* some fowls at that show. I did n't *buy* any there, I believe.

The receipts at the gates paid the expenses of the exhibition, and left a small surplus in the hands of somebody,— I never knew who,— but who took good care of the money, I have not a doubt; as most of the officers at *that* time were, like myself, "poor, but honest."

By the time this fair closed, the pulse of the "dear people" had come to be rather rapid in its throbs, and the fever was evidently on the increase. Fowls were in demand. Not *good* ones, because nothing was then said by the anxious would-be purchasers about *quality*. Nobody had got so far as that, then. They wanted *fowls* only,— hens and cocks,— to which they themselves gave a name.

Some fancied one breed, or variety, and some another; but anything that sported feathers,— from the diminutive Bantam to the stork-shaped Chinaman,— everything was being sought after by "amateurs" and "fanciers" with a zest, and a readiness to pay for, that did honor to the zeal of the youthful buyers, and a world of good to the hearts of the quiet breeders and sellers, who began *first* to get posted up, and inured to the disease.

I was an humble and modest member of this latter class. *I* kept and raised only *pure* breeds of fowls.

CHAPTER IV.

HOW "POULTRY-BOOKS" ARE MADE.

SOON after this, I learned that one Asa Rugg, of Pennsylvania (a *nom de guerre*), was in the possession of a breed of fowls that challenged all comparison for size and weight. They were called the *Chittagong* fowl, and were thus described in the poultry-books published in 1850:

"The fowl thus alluded to has been imported, within the last two or three years, into Pennsylvania, and ranks at the head of the list, in that region, for all the good qualities desirable in a domestic bird. The color is a *streaked grey*, rather than otherwise, and the portraits below" (my birds) "are fine samples of this great stock. They are designated as the GREY CHITTAGONGS."

"Asa Rugg," in his letter to me, described this stock as being at the head of the races of poultry, having "the *largest* blood in them of any variety of fowl with which he was acquainted." The pair he first sent me were light-grey and streaked, and "at less than seven months old weighed over *nineteen* pounds."

He said, in that insinuating and delicate manner so peculiar to the habits of gentlemen who possess what another wishes to buy of them,—"I did not intend, my dear Mr. B——, to part with these magnificent specimens at *any* figure whatever. I assure you I had much rather retain them; for they are *very* fine, as you would say, could you see them. If, however, you are disposed to pay my price, I shall let you have them. I really shall regret their absence from my yard, however. Try and make up your mind to be satisfied with something else — won't you? *These* fowls I must keep, if possible," &c. &c.

Now, Asa *knew* very well, if he had charged me two hundred (instead of twenty) dollars for those grey fowls, I should have taken them from him. Of course I sent for them at once; and, within ten days, they were in my poultry-house, a new wonder for the hundreds who called to see my "superb" and "extraordinary" fowls.

A competitor turned up, a few months after this, a notorious breeder in P——, who, though a respectable man, otherwise, never knew a hen from a stove-pipe, but who had more money at that time than I had, and who, in the hen-trade, possessed the impudence of the devil, without the accompanying graces to carry out his object.

This man chanced, while in Pennsylvania, to hear Asa speak of *me*, and at once he stepped in to "head me" in that quarter. He bought all the "*Grey* Chittagongs" that

Rugg had left (most of which, when they reached P——, happened to be dark *red* and *brown*), and forthwith set up an establishment in *opposition* to me; for what purpose I never knew. I did not know him from a side of sole-leather, I had never spoken to or of him, and I could not comprehend why this person should render himself, as he did, my future "death's head" in the fowl-trade.

If he went into the traffic for the purpose of making money out of it, he has found, by this time, I have no doubt, that he would have been, at the very least calculation, five thousand dollars better off had he never thrust himself into a business of which he did not know the first rudiments.

If he embarked in it to interfere with or to injure me, personally, he has now ascertained, I imagine, that it required a faster horse than *he* was in the habit of driving to keep in sight of *my* team.

If his purpose was the gratification of his own petty spite or ambition only, he has had to pay for the enjoyment of it,—ay, to his dear cost!—and he is welcome to all he ever made out of his contemptible, niggardly huckstering.

Soon after the first exhibition, it was announced by the publishers in Boston that Dr. Bennett's new Treatise would be immediately issued by them. The doctor had originally applied to the establishment in which I was then a partner, to issue this work; but I recommended him to the others, because our own facilities for getting it out were not so good as I thought were theirs.

I furnished a considerable amount of the matter for that book, and had already obtained, at my own individual expense, several of the engravings which appear in the work spoken of. After the original cuts were placed in the publishers' hands, they were reduced in size, and injured (for my purposes), as I conceived, when they finally appeared in print.

The doctor's book on poultry had been announced again and again; but it did not make its appearance in the market, in consequence of his tardiness. Week after week, and month after month, passed by, but still no Dr. Bennett's book could be found. I saw some of the proof-sheets finally, observed the fate of the illustrations of *my* fowls, and made up my mind what I would do. The book was at last announced positively to appear in three weeks.

I immediately called at a stereotype foundery, and asked how much time it would require to stereotype a work of one hundred and fifty pages for me. I was told that it would occupy three to four weeks to complete it. "Can't it be done in *one* week?" I inquired. The proprietor smiled, and said that this was impossible. I replied, "Well, sir, to-day is Tuesday. I have engaged to deliver in New York city, on the morning of a week from next Saturday, three thousand copies of a book *which I am about to write.* Is there *no* way that you can help me out?" The gentleman looked at me incredulously.

I added, "Mr. ——, I have been in the newspaper business a good many years, and I have had the message of the President of the United States — a document occupying a dozen columns of solid brevier and minion — set up and put to press within forty-two minutes from the time it reached our office. *Anything* can be accomplished, now-a-days, if we but will it."

"But, you say you *are about* to write it. When will the 'copy' be ready?" said the stereotyper.

"I have thought of this," I replied, "but a few hours. The *title*, even, is not yet decided upon. I will give you fifty pages of manuscript to-morrow morning, the next day I will add another fifty, and you shall have the whole in hand by Friday morning."

He kindly undertook to aid me. I engaged three engravers, who worked day and night upon the drawings and transfers of the fowls for my illustrations; the paper was wet down on Monday and Tuesday; I read the final revised proof of my work on Wednesday night; the book went to press on Thursday; the binders were ready for it as it came up, the covers were put on on Friday morning, and I sent to the New York house (who had bespoken them), by Harnden's Express, on Friday evening, three thousand five hundred copies of the "New England Poultry-Breeder," *illustrated with twenty-five correct engravings* of *my* choice, magnificent, superb, unapproachable, pure-bred fowls.

This book had an extraordinary sale,— far beyond my own calculations, certainly. I got it out for the purpose of "doing justice" to my own stock, and calculated that it would prove a good advertising medium for me,— which it *did*, by the way. But the demand for the "New England Poultry-Breeder" was immense. And *thirteen* different editions (varying from three thousand five hundred to one thousand copies each) were issued within as many weeks, and were sold, every copy of them. This is the true history of the "New England Poultry-Breeder."

By and by Dr. Bennett's book appeared. The market was now glutted with this kind of thing, and this work, though a good one, generally, dragged on the hands of its originators. I doubt if a thousand copies of this book ever found their way into the market, the author being too deeply engrossed with his then thriving trade, to trouble himself about urging the sale of his book, or of thinking about the interests of his publishers.

CHAPTER V.

THREATENING INDICATIONS.

ANOTHER meeting was now called at the Statehouse, which was even more fully attended than the first, and at which much more serious indications of enthusiasm were apparent.

Old men, and middle-aged farmers, and florists, and agriculturists, and live-stock breeders, from all parts of this and the neighboring states, congregated together on this eventful occasion, and entered into the debate with an earnestness worthy of so important and "glorious" a cause.

Some of the speakers had by this time got to be so elated and so ardent that they rehearsed all they knew, and some of them told of a great deal more than themselves or anybody else had ever dreamed of, bearing upon the subject of poultry-raising. But, really, the subject was an exciting one, and the talkers were excusable; they could n't help it!

Shades of morus multicaulis victims! Shadows of defunct tulip-growers! Spirits of departed Merino sheep speculators! Ghosts of dead Berkshire pig fanciers!

Where were ye all on that eventful night, when six hundred sober, "respectable" representatives of "the people" were assembled within the walls of our time-honored state edifice upon Beacon Hill, in serious and animated conclave, to decide the momentous question that "hens *was* hens," notwithstanding, nevertheless!

"Mr. President," exclaimed one of these gentlemen (whose speech was not publicly reported, I think), "Mr. President, the times is propishus. We 're a-enterin' on a new ery. The *people* is a-movin' in this 'ere great, and wonderful, and extraordinary — I may say, Mr. President, this 'ere soul-stirrin' and 'lectrefyin — branch of interestin' rural erconomy." (Applause, during which the speaker advanced a step or two nearer to the presiding officer's desk, wiped his nose fiercely upon a fiery-red bandanna handkerchief, and proceeded.)

"The world, Mr. President," he continued, "is a-growin' wiser ev'ry day,— I may say ev'ry hour, Mr. President! Ay, sir, ev'ry minute." (Loud applause, amid which one old gentleman in a bob-wig was particularly vociferous.)

"I say, Mr. President, the people is a-growin' wiser continu'lly; and by that expression, sir, I mean to convey the idee that they are a-gettin' to know more, sir! Who will gainsay this position? Whar 's the man — whar 's the er — individooal, sir — that 'll stan' up 'ere to-night, in this hallowed hall, under the shadder of this doom above our heads, sir, in view of the great American eagle yender,—that 'bird of promise,' sir,—and dispute the assertion that I now make, Mr. President, as an American citizen, without fear and without reproach!" (Deafening shouts of "Nobody! nobody *can* dispute it!")

"No, *sir!* I think not, I wot not, I ventur' not, I cal'k'late *not!* I say, Mr. President, it is no use for nun of us to contend agin the mighty ingine of progress; 'nless we 'd like to get our crowns mashed in for our pains, sir. That 's the way it 'pears to *me;* and I 've no doubt that this 'nlitened ordinance now present, sir, will agree with me on *this* p'int, and admit the truth that present indications, sir, p'int, with strikin' force, to the proberble likelihood that the deeds begun here to-night must be forever perpetooated hereafter, and that — a — they will — er — go down, sir, to our children, and our children's children, *a posteriori*, in the futur, forever!" ("Yes, yes!" and thundering applause.)

"But, sir, the p'int at issoo seems to me to be clear as the broad-faced sun on a cloudy day. I 'm no speaker, sir. I am not the man, sir, that goes about to proclaim on tops of houses! I 'm a quiet citizen, and calls myself one o' 'the people,' sir. But w'en the questions comes up of *this* natur',—w'en it 'pears to me to be so clear and so transparent, —w'en the people goes abroad, sir, in their might, and —er—and can't stay ter home,—w'en *such* things occurs, sir, then *I'm round!*" (Shouts of "Good! good! good!" the respectable old gentleman in the bob-wig creating a cloud of dust about him with his stamping and excited gestures.)

"Mr. President, I have a'most done ——" ("No, no! Go on, go on!" from all parts of the house.)

"No, sir; as I've said afore, I'm no speaker, an' I make no pretenshuns to oraterry. I'm a plain man, sir; but I feel deeply interested in this subject." (Nobody had yet ascertained what the "subject" was, because the gentleman had n't alluded to any.) "And, sir, I feel that I should be unjust to myself and to this ordinance ef I did not say what I have, sir. I go in for the poultry-breedin', sir, all over! Sir,

> I love 'em, I love 'em, — an' who shall *dar'*
> To chide me for lovin' and praisin' them *'are?*

"I love 'em, sir,—chickens or poultry,—dead or alive. My father afore me loved 'em, sir; and I'm rejoiced to see the feelin's that 's exhibited here to-night. And, 'less anybody should suspect that I have ventured upon these few remarks with mercenary motives, Mr. President (though *perhaps* no such suppersishun would animate no man's bosom), I will state, sir, that I have no fowls to sell, sir,— none whatever. *No*, sir! not a fowl! I'm a buyer, sir, — I want to *buy*," shouted the excited man,—and he sat down amid the deafening plaudits of his associates at this meeting, who fully appreciated his speech and his palpable disinterestedness.

(*Item.*—I found this gentleman the next day, and informed him that I had heard of his destitution. I had understood that he had no poultry, but was in search of *pure-blooded* stock. Before night I had fully supplied him with *genuine* samples, at thirty dollars a pair, and no "discount for cash.")

Before this meeting concluded, the prices of fowls, and eggs, and feathers, were duly discussed, the details of which I will defer to the next chapter.

But all the indications at this convention were really of a threatening character; and it would have required the strength of several stout men to have held certain of the speakers as they got warmed up, and rattled away, for dear life, upon the advantages that must accrue to the nation, in a thousand ways,-from the encouragement of this epidemic, and the certain, inevitable losses that must be sustained by "the people" if they did n't go into this thing with a rush.

Most of these speakers, however, had fowls *for sale!*

CHAPTER VI.

THE EPIDEMIC SPREADING.

WHILE all this was transpiring, my "splendid" Cochin-China fowls had arrived from England, and I had had a nice house arranged, in which to keep and exhibit them to visitors.

The pullets began to lay in January, 1850, and immediately afterwards my trade commenced in earnest, which continued, without interruption, up to the close of the year 1854.

Among the "monstrosities" presented at the second meeting at the Boston Statehouse were several propositions that were suggested by gentlemen-amateurs and farmers in regard to the price that should be fixed on, by members of the Society with the elongated title, for *eggs* sold for incubation.

One man thought that *two* dollars a dozen for most of the fancy kinds would pay well. This gentleman (I do not remember who he was) probably calculated to furnish

4

fancy eggs as a certain agricultural concern had been doing for some months: that is, by first purchasing them at a shilling a dozen from the eastern packets, or in Quincy Market. The next man thought that *three* dollars per dozen would be fair. Another member believed that *one* dollar was enough for twelve eggs, "but he did n't know much about it," he acknowledged; which was pretty evident from his remarks. At any rate, he had never fed a "laying hen" long enough on good corn to ascertain how much she would devour while she was furnishing him with the said twelve eggs, I imagine! One gentleman, more liberally disposed, probably, ventured to express his willingness to pay *five* dollars a dozen for what he wanted. I understood he got home safely after the meeting, though it was feared he would be mobbed for his temerity in making this ridiculous offer!

I had already fixed *my* price for the eggs that were to be dropped by my "extraordinary and superb" Cochin-China fowls, which by this time had got to be "the admiration of the State" (so the newspapers said). I had the *best* fowls in this world, or in any other; this being conceded by every one who saw them, there was no necessity of "talking the subject up" to anybody. I charged *twelve* dollars a dozen for my eggs — and never winked at it!

And why should n't I have the highest price? Were not my fowls the "choicest specimens" ever seen in America?

Did n't everybody so declare? Did n't the press and the poultry-books concede this, without an exception? Well, they did! And so, for months, I obtained one dollar each for my Cochin-China fowls' eggs; and I received order after order, and remittance after remittance, for eggs (at this figure), which I could not *begin* to supply.

And I did n't laugh, either! I had no leisure to laugh. I filled the orders as they came,— "first come, first served," — and for several months I found my list of promises six or eight weeks in advance of my ability to meet them with *genuine* eggs.

I was not so well informed, then, as I was afterwards. I think all the eggs that were then wanted *might* have been had. But, as the boy said, when asked where all the stolen peaches he had eaten were gone, "I donno!"

Will it be credited that, during the summer of 1850, I had dozens of full-grown men — gentlemen — but enthusiastic hen-fanciers (who had contracted the fever suddenly), who came to my residence for Cochin-China eggs, at one dollar each. and who, upon being informed that I had n't one in the house, would quietly sit down in my parlor and wait two, three, or four hours at a time, *for the hens to lay them a few*, that they might take them away with them? Such is the fact, however it may be doubted.

I subsequently sold the eggs at ten dollars a dozen; then at six dollars; and finally, the third and fourth years, at five dollars. This paid me, because I sent off a great many.

But they did n't hatch well after having been transported away and shaken over in the hands of careless and ignorant or reckless express agents. Thus the buyers came again. Many of the early fanciers tried this experiment over and over again, but with similar ill-success; and when they had expended ten, twenty, or thirty dollars, perhaps, for eggs, they would begin at the *beginning* aright, and purchase a few chickens to rear, from which they could finally procure their own eggs, and go forward more successfully. But all this took time to bring it about.

And meanwhile *somebody* (I don't say who) was "feathering a certain nest" as rapidly as a course of high-minded and honorable dealing with his fellow-men would permit.

CHAPTER VII.

ALARMING DEMONSTRATIONS.

My premises were literally besieged with visitors, and my family attendants were worn out with answering the door-bell summons, from morning till night.

"Is Mr. B—— at home? Can we see his Cochin-Chinas? Can we look at Mr. B——'s fowls? Might we take a look at the chickens?" were the questions from sun to sun again, almost; and I was absolutely compelled, in self-defence, to send the fowls away from home, for a while, for the sake of relief from the continual annoyances to which, in consequence of having them in my yard, I was subjected.

Fifteen, twenty, often forty callers in a single day, would come to see my "magnificent" Cochin-China fowls. But I sent them off, and then "the people" cried for them!

"Who's dead?" queried a stranger, passing my door one day, and observing the carriages and vehicles standing in a line along the front of my garden-fence.

"Nobody, I guess," said another; "that's where the *Cochin-Chinas* are kept."

"The what?"

"The Cochin-Chinas."

"What 's them?"

"Don't you know?"

"No; never heard of 'em, afore."

"Never heard of Burnham's Cochin-Chinas?"

"Never! What are they?"

"Well, I reckon you ha'n't lived in these 'ere parts long, my friend," continued the other; "and you 'd better step in and look at 'em."

In came the stranger, and after examining the fowls he returned.

"How do you like 'em?" asked the man who had already seen them, and was waiting for his friend outside.

"They 're *ronchers*, that 's a fact!" exclaimed the gratified stranger. And this was the universal opinion.

Nobody had ever seen such fowls (*I* had seen a good many better ones!) — nobody had ever beheld any so large, so heavy, so fine. And every one who came to look at them purchased or engaged either eggs or chickens from these "extraordinary" and "never-to-be-too-much-lauded" royal Cochin-China fowls!

For my first broods of chickens (at three and four months old) I readily obtained twenty-five dollars a pair; and every one of them went off "like hot cakes" at this figure. It was too low for them, altogether; and I had occasion to regret, subsequently, that I did not charge fifty dollars a pair; — a price which I might just as easily then have obtained as if I had charged but one dollar a pair, as events proved to my satisfaction.

But everything connected with this fever could not well be learned at once. I was not a very dull scholar, and I progressed gradually. One year after the receipt of my Cochins, I got my own price for them, ask what I might. I sold a good many pairs at one hundred dollars the couple; and, oftentimes, I received this sum for a trio of them.

Things begun to look up with me. I had got a very handsome-looking stock on hand, at last; and when my numerous customers came to see me, they were surprised (and so was *I*) to meet with such "noble" samples of domestic fowls. "Magnificent!" "Astonishing!" cried everybody.

A splendid open carriage halted before my door, one day, and there alighted from it a fine, portly-looking man, whom I had never seen before, and whose name I did not then learn; who, leaving an elegantly-dressed lady behind in the vehicle, called for me.

I saw and recognized the *carriage*, however, as one of Niles'; and I was satisfied that it came from the Tremont House. As soon as the gentleman spoke, I was also satisfied, from his manner of speech, that he was a Southerner. He was polite and frank, apparently. I invited him in, and he went to look at my fowls; that being the object, he said, ot his visit.

He examined them all, and said, quietly:

"I'd like to get half a dozen of these, if they did n't come too high; but I understand you fanciers have got the price up. I used to buy these chickens for a dollar apiece. *Now*, they say, you 're asking five dollars each for them."

I showed him my stock,—the "*pure*-bred" ones,—and informed him at once that I had not sold any of *my* chickens, latterly, at less than *forty* dollars a pair.

He was astounded. He did n't want any—much: that is, he was n't particular. He could buy them for five dollars; should n't pay that, *no*how; wanted them for his boy; would come again, and see about it, &c. &c.

A five-year-old stag mounted the low fence at this moment, and sent forth an electrifying crow, such as would (at that period) have taken a novice "right out of his boots;" and a beautiful eight-pound pullet showed herself beside him at the same time. The stranger turned round, and said:

"There! What is your price for such a pair as that, for instance?"

"Not for sale, sir."

"But you *will* sell them, I s'pose?"

"No, sir. I have younger ones to dispose of; but *that* pair are my models. I can't sell *them*."

The gentleman's eye was exactly filled with this pair of chickens.

"What will you *take* for those two fowls?"

"One hundred dollars, sir," I replied.

"I guess you will — when you can get it," he added.— "Name your lowest price, now, for those two. I want *good* ones, if any."

"I prefer to keep them, rather than to part with them at *any* price," I insisted. "If, however, a gentleman like yourself, who evidently knows what good fowls are, desires to procure the choicest specimens in the country, why, I confess to you that those are the persons into whose hands I prefer that my best stock should fall. But I will show you some at a lower figure," I continued, driving this pair from the fence.

"Don't you! Don't drive 'em away!" said the gentleman; — "let 's see. That 's the cock?"

"Yes, sir."

"And this is the hen?"

"Yes."

"One *hundred* dollars! You don't *mean* this, of course," he persisted.

"No, I mean that I would rather keep them, sir."

"Well — I 'll ——— *take them*," said the stranger. "It 's cruel. But, I 'll take them;" and he paid me five twenty-dollar gold pieces down on the spot, for two ten-months-old chickens, from my "splendid" Royal Cochin-China fowls.

He had a tender spot *somewhere*, that I had hit, during the conversation, I presume. He took the two chickens into his carriage, and I have never seen or heard from him from that day to this. I trust, however, if "these few lines" should ever meet his eye, that his poultry turned out well, and that he himself is in good health and spirits!

I called this gallant young cock "Frank Pierce," in honor of my valiant friend now of the White House, at Washington. It will be seen that I thus sold Frank for fifty dollars; a sum which the majority of the people of this country have since most emphatically determined was *a good deal more than he ever was worth!*

CHAPTER VIII.

THE FEVER WORKING.

About this time an ex-member of Congress, formerly from Pennsylvania, was invited to deliver the address before one of the county agricultural societies of that state (where the fever had now begun to spread with alarming rapidity), who, in the course of his speech on that occasion, delivered himself of the following pointed and forcible remark.

Speaking of poultry and the rare qualities of certain domestic fowls, he said, "Ladies and gentlemen, next to a beautiful woman, and an honest farmer, I deem a Shanghae cock the noblest work of God!"

Now, this expression might be looked upon, by some persons, as savoring of demagogism, or, at the least, as an approach to "running this thing into the ground" (or into the air); but the honorable gentleman no doubt felt just what he said. I have seen many sensible men who felt worse than this — a good deal — on this self-same subject; and who expressed themselves much more warmly in regard to the characteristics and beauties of domestic poultry; but, to be sure, it was *after* they had "gone through the mill," and had come out at the *small* end of the funnel.

In New England, especially, prior to the *second* show of poultry in Boston, the fever had got well up to "concert pitch;" and in New York State "the people" were getting to be very comfortably interested in the subject — where *my* stock, by this time, had come to be pretty extensively known.

The expenses attendant upon this part of the business, to wit, the process of furnishing the requisite amount of information for "the people" (on a subject of such manifestly great importance), the *quantum sufficit* in the way of drawings, pictures, advertisements, puffings, etc., through the medium of the press, can be *imagined*, not described.

The cost of the drawings and engravings which I had executed for the press, from time to time, during the years 1850, '51, '52, and '53, exceeded over eight hundred dollars; but this, with the descriptions of my "rare" stock (which I usually furnished the papers, accompanying the cuts), was *my* chosen mode of advertising. And I take this method publicly to acknowledge my indebtedness to the press for the kindness with which I was almost uniformly treated, while I was thus seriously affected by the epidemic which destroyed so many older and graver men than myself; though few who survived the attack "suffered" more seriously than I did, during the course of the fever. For instance, the large picture of the fowls which I had the pleasure of sending to Her Majesty Queen Victoria (in 1852), and which appeared in *Gleason's Pictorial*, the *New York Spirit of the Times*, *New England Cultivator*, &c., cost me, for the original drawing, engraving, electrotyping, and duplicating, *eighty-three dollars.*

All these expenses were cheerfully paid, however, because I found my reward in the consciousness that I performed the duty I owed to my fellow-men, by thus aiding (in my humble way) in disseminating the information which "the people" were at that time so ravenously in search of, namely, as to the person of whom they could obtain (without regard to price) the *best* fowls in the country.

This was what "the people" wanted; and thus the malady extended far and wide, and when the fall of 1850 arrived, buyers had got to be as plenty as blackberries in August, whilst sellers "of reputation" were, like the visits of angels, few and far between. *I* was, by this time, considered "one of 'em." I strove, however, to carry my honors with Christian meekness and forbearance, and with that becoming consideration for the wants and the wishes of my fellow-men that rendered myself and my "purely-bred stock" so universally popular.

Ah! when I look back on the past,— when I reflect upon the noble generosity and disinterestedness that characterized all my transactions at that flush period,— when I think of what I did for "the cause," and how liberally I was rewarded for my candor, my honesty of purpose, and my disingenuousness,— tears of gratitude and wonder rush to my eyes, and my overcharged heart only finds its solace by turning to my ledger and reading over, again and again, the list of prices that were then paid me by "the people," week after week, and month after month, for my "magnificent samples" of "pure-bred" Cochin-China chickens, the original of which I had imported, and which were *said* to have been bred from the stock of the Queen of Great Britain.

But, the Mutual Admiration —— I mean, the "Society" whose name was like

"Lengthened sweetness, long drawn out,"

was about to hold its second annual exhibition; and, as the number of its members had largely increased, and as each and all of those who pulled the wires of this concern (while at the same time they were pulling the wool over the eyes of "the people") had plans of their own in reference to details, I made up my mind, although I felt big enough to stand up even in this huge hornet's nest of competition, to have things to suit *my* "notions."

I *now* had fowls to sell! I had raised a large quantity of chickens; winter was approaching, corn was high, they required shelter, the *roup* had destroyed scores of fowls for my neighbors, and I did n't care to winter over three or four hundred of these "splendid" and "mammoth" specimens of ornithology, each one of which could very cleverly dispose of more grain, in the same number of months, than would serve to keep one of my heifers in tolerable trim.

Such restrictions were proposed by the officers of the Society with the lengthened cognomen, that my naturally democratic disposition revolted against the arbitrary measures talked of, and I resolved to get up an exhibition of my own, where this matter could be talked over at leisure, and which I did not doubt would "turn an honest penny" into my own pocket; where, though I had done *well* thus far, there was still room, as there was in hungry Oliver Twist's belly, for "more."

CHAPTER IX.

THE SECOND POULTRY-SHOW IN BOSTON.

On the 2d, 3d, and 4th days of October, in the year of our Lord 1850, the "grand exhibition" (so the *Report* termed it), for that year, came off at the large hall over the Fitchburg Railroad Dépôt, in Boston, "which proved a most extensive and inviting one" (so continued the Report), "far exceeding, both in *numbers* and in the *quality* of specimens offered, anything of its kind ever got up in America.

"The birds looked remarkably fine in every respect, and the undertaking was very successful. A magnificent show of the feathered tribe greeted the thousands of visitors who called at the hall, and all parties expressed their satisfaction at the proceedings.

"The Committee awarded to George P. Burnham, of Melrose, the *first* premiums for fowls and chickens. The prize birds were the '*Royal Cochin-Chinas*' and their progeny, which have been bred with care from his imported stock; and which were generally acknowledged at the head of the list of specimens."

The prices obtained at this exhibition ranged very high, and "full houses" were constantly in attendance, day and evening, to examine and select and purchase from the "pure-bred" stock there. "Mr. Burnham, of Melrose" (continued the Report), "declined an offer of $120 for his twelve premium Cochin-China chickens, and subsequently refused $20 for the choice of the pullets."

"The show was much larger than the first one, and the character of the birds exhibited was altogether finer, though the old fowls were, for the most part, moulting. A deep interest was manifested in this enterprise, and it went off with satisfaction to all concerned," added the Report.

In order that the details of this experiment (which *I* projected and carried through, myself) may be appreciated and understood, I extract from the "official" Report the following items regarding this show, the expenses, the prize-takers, &c.

The "Committee of Judges," consisting of myself, G. P. Burnham, Esq., and a gentleman of Melrose, made the following statements and "observations," in the *Report* above referred to:

"The Exhibition was visited by full ten thousand persons, during the three days mentioned. The amount of money received for tickets was four hundred and seventy-three dollars and thirty-eight cents; and the following disbursements were made:

Cash paid for		rent of hall,	$175.00
"	"	amount of premiums and gratuities,	135.00
"	"	for lumber and use of tables,	17.60
"	"	for lighting hall, advertising, etc.,	70.40
"	"	tickets, cards and handbills,	18.21
"	"	carpenters and attendants,	27.50
"	"	police and door-keepers,	15.00
"	"	grain, seed, buckets, pans, etc.,	25.56
"	"	coops, cartage and sundries,	7.37
		Total expenses,	$491.64
		Amount received, as stated,	473.38
		Deficit,	$18.26"

When the state of the funds was subsequently more particularly inquired into, however, it was found that the amount of money actually received at the door was a little rising *nine* hundred and seventy dollars, instead of "*four* hundred and seventy-three," as above quoted. But this was a trifling matter; since the "Committee of Judges" spoken of above accounted for this sum, duly, in the final settlement.

The "Committee" aforesaid awarded the following premiums at this show, after attending to the examination confided to them — namely:

"*First* premium, for the best six fowls contributed, to *George P. Burnham*, of Melrose, Mass., $10.

"For the three best Cochin-China Fowls (Royal), to *George P. Burnham*, Melrose, Mass., $5.

For the twelve best chickens, of this year's growth (Royal Cochin-China), to *George P. Burnham*, Melrose, $5."

And there were some *other* premiums awarded, I believe, there, but by which I was not particularly benefited; and so I pass by this matter without further remark, entertaining no doubt whatever that all the gentlemen who were awarded premiums (and who obtained the amount of the awards) exhibited at the Fitchburg Hall Show *pure*-bred fowls.

After making these awards, the "Committee of Judges" (consisting, as aforesaid, of myself, Mr. Burnham, and a fancier from Melrose) state that "they find great pleasure"—(mark this!)—"they find great pleasure in alluding again to the splendid contributions" of some of the gentlemen who had fowls in this show,—and then the Report continues as follows:

"The magnificent samples of *Cochin-China* fowls, contributed by G. P. Burnham, of Melrose, were the theme of much comment and deserved praise. These birds include his imported fowls and their progeny — of which he exhibited nineteen splendid specimens. To this stock the Committee unanimously awarded the *first* premiums for fowls and chickens; and finer samples of domestic birds will rarely be found in this country. They are bred from the Queen's variety, obtained by Mr. Burnham last winter, at heavy cost, through J. Joseph Nolan, Esq., of Dublin, and are unquestionably, at this time, the finest thorough-bred Cochin-Chinas in America."

My early hen-friend the "Doctor" — alluded to in the opening chapter of this book — exhibited a fowl which the "Committee" thus described in their report:

"The rare and beautiful imported *Wild India Game* hen, contributed by Mr. B. F. Griggs, Columbus, Geo., was a curiosity much admired. This fowl (lately sold by Dr. J. C. Bennett, of Plymouth, to Mr. Griggs, for $120) is thorough game, without doubt; and her progeny, exhibited by Dr. Bennett, were very beautiful specimens. To this bird, and the '*Yankee Games*' of Dr. Bennett, the Committee awarded a gratuity of $5."

So miserable a *hum* as this was, I never met with, in all my long *Shanghae* experience. It out-bothered the Doctor's famous "Bother'ems," and really out-*Cochined* even my noted Cochin-Chinas! But I was content. *I* was one of the "Committee of Judges." I had forgot!

This Committee's Report was thus closed:

"It has been the aim of the Committee to do *justice* to all who have taken an interest in the late Fowl Exhibition, and they congratulate the gentlemen who have sustained this enterprise upon its success."

They did *ample* justice to this Wild Bengal Injun Hen, that is certain. The Cochin-China trade received an impulse (after this show concluded) that astonished even *me*, and I am not easily disturbed in this traffic. And I have no doubt that the people who paid their money to witness this never-to-be-forgotten (by me) exhibition, were also satisfied.

The experiment was perfectly successful, however, throughout. I forwarded to all my patrons and friends copies of this Report, beautifully illustrated; and the orders for "*pure*-bred chickens from the *premium* stock" rushed in upon me, for the next four or five months, with renewed vigor and spirit.

This first exhibition at the Fitchburg Dépôt Hall proved to me a satisfactorily profitable advertisement, as I carried away all the premiums there that were of any value to anybody. But then it will be observed that the "Committee of Judges" of this show were my "friends." And, at that time, the competition had got to be such that all the dealers acted upon the general democratic principle of going "for the greatest good of the greatest number." In my case, I considered the "greatest number" Number *One!*

CHAPTER X.

THE MUTUAL ADMIRATION SOCIETY'S SECOND SHOW.

In the month following, to wit, on the 12th, 13th and 14th of November, 1850, the second annual exhibition of the Simon Pure Society with the extended title was held at the Public Garden, in Boston.

No premiums were offered by the society this year, and there was n't much to labor for. I was a contributor, and I believe I was elected a member of the Committee of Judges that year. How, I did not know. At any rate, I wrote the published *Report* upon the exhibition. A Mr. Sanford Howard was chairman of this committee, if I remember rightly; and though undoubtedly a very respectable and well-meaning man (if he had not been so, he would n't have been placed on a Committee of Judges with *me*, I imagine), this Mr. Howard knew positively *nothing* whatever in regard to the merits or faults of poultry generally. He had acquired some vague notions about what he was pleased to term "crested" fowls, and five-toed, white-legged, white-plumed, white-billed, white-bellied Dorkings,— of which he conversed technically and learnedly; but as to his knowledge of the different varieties and breeds of domestic poultry then current, and their characteristics, it was evidently warped and very limited.

But Mr. Howard had been connected for some months with a small monthly publication in New York State, and, like myself, I presume, among the board (God knows who they were, but *I* don't, and never did!) who originally chose this "Committee," he had "a friend at court," and was made *chairman* of the committee too,— *how*, I never knew, either.

In their Report, the Committee observe, again, that "*never* in this country, if in the world, was there collected together so large a number of domestic fowls and birds as were sent to this exhibition, probably; and, though the most liberal arrangements were made in advance, it was found that the accommodations, calculated for *ten thousand specimens*, were entirely insufficient. The Committee merely allude to this fact to show the actual extent of this enterprise, and the importance which the undertaking has assumed, in a single year from the birth of the Association.

"According to the records of the Secretary, there were contributed to the Society's exhibition of 1850 some four hundred and eighty coops and cages. There were in all over three hundred and fifty contributors; in addition to which about forty coops, containing some six hundred fowls, were sent to the Garden and received on exhibition upon the two last days of the Show; and which could not be recorded agreeably with the regulations made originally.

"The palpable improvement in the appearance of the fowls exhibited in 1850, as compared with the samples shown in 1849, offers ample encouragement to breeders for *further and more extended efforts;* and your Committee would urge it upon those who have already shown themselves competent to do so much, *to go on and effect still greater progress* in the improvement of the poultry of New England."

This Report (the second of the series) did *my* stock ample justice, I have not a doubt. I wrote it myself, and intended that it should do so. The text was in nowise changed when printed, and a reference to the document (for that year) will convince the skeptical—if any exist—whether I was or was not acquainted with adjectives in the superlative degree!

A very singular occurrence took place about this time, the *basis* of which I did not then, and have never since, been able to comprehend, upon any principles of philosophy, economy, business, benevolence, or even of sanity. But I am not very clear-headed.

In the *addenda* to my Report (above named) there appeared the anexed statement, by somebody:

"The Trustees refer to the following with mixed pride and pleasure; the munificence and motive of the gift are most creditable. A voluntary kindness such as that of Mr. Smith is a very gratifying proof that the labors of the Society are not regarded by enlightened men as vain:

"*Boston*, 12*th February*, 1851.

"G. W. SMITH, ESQ.

"SIR: A meeting of the Trustees of the 'New England Society for the Improvement of Domestic Poultry' was held last evening, Col. Samuel Jaques, President of the Society, in the chair, and a full quorum being present, when the Treasurer announced the receipt of your very handsome *donation of one hundred and fifty dollars* in aid of the Society's funds; whereupon it was moved, and unanimously agreed, that the most grateful thanks of the Society were justly due to you for such a munificent testimony of your desire for its prosperity; that the Secretary communicate to you the assurance of the high appreciation with which the donation was received; and that its receipt, and also a thankful expression of gratitude towards you, should be placed on the records of the Society.

"I can only reiterate the sentiments contained in my instructions, in which I fully and gratefully concur; and, with best wishes for your long-continued welfare,

"I am, sir, very truly yours,

"JOHN C. MOORE, *Rec. Secretary*."

Now, it will be observed that this was not *John* Smith who presented this sum, but another gentleman, and a different sort of individual altogether. He gave it (one hundred and fifty dollars in hard cash) the full value of a nice pair of my *best* "pure-bred" Cochin-Chinas, without flinching, without any fuss, outright, freely, "in aid of the Society's funds."

Liberal, generous, benevolent, charitable, kindly Mr. Smith! You did yourself honor! *You* were one of the kind of men that I should very much liked to have had for a customer, about those days. But, after due inquiry, I ascertained that you did not keep, or breed, poultry. You were only a "friend" to the Society with the elongated name,—the *only* friend, by the way, it ever had! Heaven will reward you, Mr. Smith, sooner or later, for your disinterestedness, but the Society never can. Be patient, however, and console yourself with the reflection that he who giveth to the poor, lendeth, &c. &c. The Society with the long-winded title was *poor* enough, and you cannot have forgotten that he who casteth his bread (or money) upon the waters will find it, after many days. You will find yours again, I have no doubt; but it will be emphatically "after *many* days."

The second show closed, the expenses of which reached the sum of one thousand and twenty-seven dollars eighteen cents, and the receipts at which amounted to one thousand and seventy-nine dollars eighty-four cents, exclusive of the above-named donation. The Society had now a balance of two hundred and two dollars sixty-six cents in hand, and it went on its way rejoicing.

THE SHANGHAE REFERRED TO IN LETTER NO. 17. — (See page 80.)

Col. Jaques (the first President) now "resigned his commission," and Moses Kimball, Esq., was chosen in his stead. I found myself once more among the Vice Presidents, John C. Moore was elected Secretary, Dr. Eben Wight was made Chairman of the Board of Trustees, and H. L. Devereux became Treasurer for the succeeding year.

These officers were all "honorable men," who were thus placed in position *to watch each other!* The delightful consequences can readily be fancied. What my own duties were (as Vice-President) I never knew. I supposed, however, that, as "one of 'em" thus elevated in official rank, I was expected to do my uttermost to keep the bubble floating, and to aid, in my humble way, to maintain the inflation. And I acted accordingly; performing my duty "as I understood it"!

CHAPTER XI.

PROGRESS OF THE MALADY.

IMMEDIATELY after this second exhibition, the sales of poultry largely increased. Everybody had now got fairly under weigh in the hen-trade; and in every town, at every corner, the pedestrian tumbled over either a fowl-raiser or some huge specimen of unnameable monster in chicken shape.

I had been busy, and had added largely to my "superior" stock of "pure-blooded" birds, by importations from Calcutta, Hong-Kong, Canton and Shanghae, direct. In two instances I sent out for them expressly, and in two or three other instances I had obtained them directly from on shipboard, as vessels arrived into Boston and New York harbors.

I was then an officer in the Boston Custom-house,—a democrat under a whig collector,—otherwise, a live skinned eel in a hot frying-pan. But I found that my business had got to be such that I could not fulfil my duty to Uncle Sam and attend appropriately to what had now got to be of very much greater importance to me; and so I resigned my situation as Permit Clerk at the public stores, very much to the regret of everybody in and out of the Custom-house, and especially those who were applicants for my place!

I had purchased a pretty estate in Melrose, and now I enlarged my premises, added to my stock, and raised (during the summer and fall of 1851) over a thousand fowls, upon my premises. This did not begin to supply the demands of my customers, however, or even approach it. And, to give an idea of my trade at that period, I will here quote a letter from one of my new patrons. It came from the interior of Louisiana, in the fall of 1851.

"GEO. P. BURNHAM, ESQ., BOSTON.

"I am about to embark in the raising of poultry, and I hear of yourself as an extensive breeder in this line. Do me the favor to inform me, by return mail, what you can send me *one hundred pairs* of Chinese fowls for, of the yellow, red, white, brown and black varieties; the cocks to be not less than eight to ten months old, and pullets ready to lay; say twenty pairs of each color. And also state how I shall remit you, in case your price suits me, &c.

"—— ——."

I informed this gentleman that I had just what he wanted (of course), and that if he would remit me a draft by mail for fifteen hundred dollars — though this price was really too low for them — I would forward him one hundred pairs of fowls "that would astonish him and his neighbors." Within three weeks from the date of my reply to him, I received a sight draft from the Bank of Louisiana upon the Merchants' Bank, Boston, for fifteen hundred dollars. I sent him such an invoice of fowls as pleased him, and I have no doubt he was (as he seemed to be) perfectly satisfied that he had thus made the best trade he ever consummated in the whole course of his life.

During the next spring I bred largely again, and supplied all the best fanciers in New England and New York State with stock, from which *they* bred continually during that and the succeeding year.

In the spring of 1852 the Mutual Admiration Society of hen-men got up their *third* show, at the Fitchburg Dépôt (in *May*, I think), where a goodly exhibition came off, and where there were now fowls *for sale* of every conceivable color and description, good, bad, and indifferent. I contributed as usual, and, as usual, carried away the palm for the *best* samples shown. And here was evinced some of the shifts to which certain hucksters resorted, to make "the people" believe that white was black, that they originally brought this subject before the public eye, and that *they* only possessed the pure stock then in the country.

Reverends, and doctors, and deacons, and laymen,—all were there, in force. Every man cried down every other man's fowls, while he as strenuously cried up his own. Upon one cage appeared a card vouching for the fact that a certain *original* Shanghae crower within it, all the way from the land of the Celestials, weighed fourteen pounds and three ounces, and that a hen, with him, drew nine pounds six ounces (almost twenty-four pounds). When the birds were weighed, the first drew ten and a half pounds, and the other eight and a quarter only. This memorandum appeared upon the box of a *clergyman* contributor, who had understood that size and great weight only were to be the criterion of merit and value thenceforward. Another contributor boldly declared himself to be the original holder of the only good stock in America. A third claimed to be the father of the current movement, and had a gilded vane upon his boxes which he asserted he had had upon his poultry-house for five years previously. Another stated that all my fowls (there shown) were bred from *his* stock. And still another proclaimed that the identical birds which I contributed were purchased directly of him; he knew every one of them. Finally, one competitor impudently hinted that my birds actually then belonged to *him*, and had only been *loaned* to me (for a consideration) for exhibition on this occasion!

When the fair closed, however, the matter was all set right, as may be gathered from the following extract from the official Report of the third show, of the Committee of Judges, of which I was *not* a member:

"At this third Boston Show," says the Committee, "the best and most faultless descriptions of Red and Buff Shanghaes were shown by G. P. Burnham, Esq., and others. And of the Cochin-Chinas, the specimens of Geo. P. Burnham, etc., were each and all notable, and worthy of public appreciation."

This was satisfactory to me, and I made the most of this "werry fav'rable opinion" of the august Committee,— who added the following, in their Report, in reference to the action of Southern purchasers:

"It seems, from reliable information received by members of the Committee, that fowls raised in New England, and exported South, attain to a much larger size, and are vastly more prolific, than in our colder climate. This is specially so in reference to the produce of stocks recently imported from the East, namely, the Shanghaes, Cochin-China fowls, and others of larger varieties. *So sensible have some of the most eminent Southern breeders become that such is the case, that they are annually in the habit of buying their young stock from the Northern States, and they find the system profitable.* In this way, New England bids fair to become the supply-market, in a great measure, for the South and West."

This was beautiful! "*Annually* in the habit." I liked *that* portion of it. And Southern buyers seemed to like it, too, judging from the manner in which orders poured in upon us, after this gentle hint from *such* authority! I believe that the Chinese fowls really did better in the South than they did with us, this way. At least, I *hope* they did!

CHAPTER XII.

MY CORRESPONDENCE.

By this time my correspondence with gentlemen in all parts of America and Great Britain had got to be rather extended. I took from the post-office from ten to twenty-five or thirty letters, daily; and amongst them were some curious samples of orthography, etymology, syntax, and prosody. I offer the annexed specimens — of course without names or dates — merely to show how the young aspirants for fame (in the poultry-trade) felt, about those days; and, also, to give some idea of the progress of the fever among us, as time passed by, etc. etc.

No. 1.

Sir — Mr. Burnham;

i red in Nu england poultry breeder that yu kep fouls an aigs for sail. i want one duzen aigs if tha doant cum tu tu mutch. ime a poor mann an carnt pa a gret pris. wot can yu cend me a duzen of yure best aigs for. ansur by male and direck yure leter tu me tu mi dress.

Yr Respec'y, &c.

—— ——.

No. 2.

My Dear Sir:

I am a poor clergyman, and I have some leisure, which I can devote to raising a few good fowls. If your price is not too high for the rather limited contents of my purse, please inform me, by return of mail, what you can furnish me with *pure* Cochin-China eggs for. I am desirous to procure a few; and I prefer that you would select for me,—in a half-dozen, say two *male* and four *female* eggs. I suggest this, because I am informed that your long experience in this interesting branch of rural economy has enabled you to decide (upon examining them) whether eggs will produce cocks or pullets. Your early answer will confer a favor on, Sir, yours, truly,

—— — ——.

No. 3.

Mr. Burman:

I close you ten dolls. Cend me a doz. of your Cotchen Chiny eggs rite away — cause I hav a hen thats been a setting on some stones I put under her now most a week. You rote me that you would hav them about this time, you know. Cend them by ——'s Express, and tell the man who fetches them not to turn the box over, at all. I want half and half — that is to say, half cock eggs, and half hen eggs. You know what I mean by this. Them that has the sharp ends on to one side — them 's the cocks, and them that 's round and smooth at both ends — them 's the hens. Forwud immediately, and mark *with care glass this side up — don't shake this with speed.*

Yours, &c.

—— ——.

No. 4.

G. P. Burnham, Esq.

Dear Sir: I saw your beautiful Cochin-China fowls last week, in the paper, and am desirous to obtain a few eggs from them, if possible.

Will they hatch under our common hens? Or, must we have the *pure* bloods to sit upon them? I am a novice, somewhat, in this business. I enclose you twelve dollars (the price for a dozen, I believe), which please forward, at your early convenience, by express, and oblige

Yours, &c.

—— — ——.

No. 5.

Friend Burnham:

Enclosed please find ten dollars for another dozen of your *pure* Cochin-China eggs. The first ones you sent me (from some cause) did not hatch. I have kept a hen (a very good sitter, too) upon that first lot, *constantly*, for four weeks, now — and I don't believe I shall get a chick, you see! So, please forward these now, *right away* — because my hen will get tired of waiting, you know, if I don't keep her right along, steady. The $10 you will find within. Yours, resp'y,

—— ——.

P. S. Can you inform me what is good for *lice* on fowls? I find that my hen is covered with a million of them, now. Don't forget this, please.*

No. 6.

SUR — wen i cum to boston nex weak i want to see yure poltry i am a ole hand at the bizness myself an I like to see good kinds of poltry every ware. i see yurn in the paper an i like them verry much can yu sel a hen without a cock, i have sevral cocks now of the *black dawkin* pure bred and fine an i would change one of them with yu for a cochon chiner hen if yu say so. answer by fust male.

Yure in haist

—— ——.

Mr. P. G. Barnum,
boston.

* After a hen had set over four weeks on her nest, I should suppose she *might* have been thus affected!

No. 7.

Dear Sir: Yours duly received. I did not suppose that the price of the "Cochins" was so high — but I must have a trio of them, at *any* figure. I enclose you fifty dollars for them, agreeably with your proposal, relying upon your known good taste in selections, and upon your proverbial reputation as regards the keeping only of *pure* stock. Send them by Adams & Co.'s Express, in a roomy cage. If they are prime, my neighbors will very shortly order from you, I am sure.

Yours, resp'y,

—— — ——.

No. 8.

Mr. Barnam:

Them two fowls I bought of you, by seeing the pictur in the newspaper, and which I paid you $35 cash down on the nail for, aint what they 're cracked up to be — not by a long short, sir. Now, what I want you to do is to sen me back my munney, or I 'll prosecute you and put you in prizon for cheating people by false pertences. I was so mad when I took them out of the box that I 'd a good mine to kill an eat em both on the spot.* I aint no *hen*-man, I 'd have you to understan, an you can't come none of this kine of nonsense over *me*. Sen me back my munney, or I 'll complain of you in tu days before a Justis of

* O, the cannibal!

the Peas — a friend of mine, that 'll give you *fits* if you *air* a big man. I don't keer for that. I want my munney. The fowls is both sick, too. Answer this tu once, or els sen me back my munney.*

—— — ——.

No. 9.

G. P. BURNHAM, ESQ.:

I saw a cage of superb Cochin-China fowls from your yard, yesterday, *en route* to Mobile. Can you duplicate them? If so, at what price? I had understood that a Mr. —— kept choice fowls. I visited his place, but saw none there that seemed worth the taking away. If you can send me such a trio as I saw at Adams & Co.'s, let me know it immediately, and your price for them. How shall I remit you? Yours, &c.

—— ——.

No. 10.

MR. BURNHAM:

I enclose you one hundred dollars, by check on Shoe and Leather Dealers' Bank, Boston (No. 417), to your order, for the fine fowls you describe in yours received this day. They should be *good* ones, as I have no doubt they are. Forward, at once,

And believe me,

Yours,

—— — ——

* I never heard from this customer again, and should now be glad to know if he ever got his "munney"!

No. 11.

G. B. Burnham, Boston.

Sir: When I paid you $25 (twenty-five dolls.) for a pair of *Cochin-China* chickens, according to your own terms, I did not suppose you would dare to send to *me* (whom you must know to be a judge of all kinds of poultry) a pair of *Shanghaes*, instead of those I ordered ! * I want none but *pure*-bred fowls in my collection, nor will I have them there, either. I have now a plenty of the Shanghaes, and I ordered a pair of Cochin-Chinas of you. Now, I want to know what you will do in this matter. Will you send me a pair of *Cochins*, or not? That is all I want to know at present. From

Yours, truly,

—— ——.

P. S. I am a lawyer by profession; and I submit to no imposition of this sort, you may be sure.

No. 12.

G. P. Burnham.

My Dear Sir:

The magnificent "Cochin-China" birds you forwarded me last are the admiration of every one who beholds them; and I am greatly your debtor for this superb lot of fowls. My neighbor, Hon. Mr. M——, desires me to request you to forward him four as nearly like mine as

* *Here* was a "lawyer," who knew the difference between a Cochin-China and a Shanghae !

possible, and your draft on me, at sight, for the cost, will be duly honored. He can afford (and is willing) to pay liberally for them.* Charge him accordingly; but be careful that you do not send him finer samples than *mine* are, — which, by the way, I do not think possible. I enclose you draft for $120, on Merchants' Bank, Boston, for your bill. And am Yours, truly, —— ——.

No. 13.

SIR — I hav alwas heerd yu was a scamp, and now I *know* yu are.† Them egs yu sent me was smasht all up, an they was runnin' down the sides of the box. What am I to do with them, sir — do yu think? Do yu spose I 've gut money so plenty as to throw it way in this manner? Yu didnt put in *harf* meal anuf, and the hole of them was spilte, besides being roten I hav no manner of dout. Now if yu send me back the six dolls. that the postmaster see me put into my fust letter to yu, all 's well an good. And ef yu don't, see if I don't publis yu and yure caracter tu the hole wurld yu infermus cheet yu. Yu'd aughter be ashamed tu send a man egs that wa, anny how. So no more at present tell I heer from yu. —— ——.

No. 14.

FRIEND BURNHAM:

I have heard creditable accounts of thy poultry (of

* This was the kind of gentleman I loved to fall in with.

† *Some* persons would consider this personal!

the Cochin-China variety), and I am induced from common rumor to believe thee a man who dealeth justly and honorably. I desire to procure a few of these choice fowls, if not too expensive; and will thank thee to inform me what thy price is for such, at ages varying from four to eight months old. Thy early reply will oblige thy friend and well-wisher,

—— —— ——.

No. 15.

G. P. Burnham, Esq. — Dear Sir: Send me ten trios more of the Cochin-China chickens, *immediately*. If you can put them down to $35 the trio, now, it will leave me a better margin. All the others are sold, at $60 the trio. Enclosed is draft on Bank of Commerce, Boston, for $400. In haste, yours,

—— ——.

No. 16.

Sir —

I want tu get sum coshin chiney aggs, them as will hatch out chickns with fethers onto the leggs an no mistaik. if you got them kind yu can cend me wun dusen an i will cen yu bak the munny wen the chickns is hached with fethers onto there leggs not otherwise. If yu dont like tu cend them on this turms yu can keepe 'em yureself. I bort too duzsen eg in bostun an their wasnt none of em had no fethers on the leg, i mene the chick'ns, wen tha was hached. an I dont expek i shall be fuled no mor by no such humbugg by a good dele. i pade my munny for genwine aigs and I donte see no reesun wy peeple isn't onnest. How could i tell wether their was chickns in the egs or not? of course i cou'dn't. and i doant consider sech bissiness no bettern than cheetin rite out. i bort em *twict* this wa, an i sharnt be fuled agin arter waitin as I did both times over three weeks. ef yu will plese to sen me the pure aigs abuv menciond and wate tell tha hach fether leggs chickns, well an good. ive no dout yu air a onnest man, cos all the noospapers pufs yu. But sum of the hen traiders aint no better than thaid oughter be — that's *my* pinion.*

Yours &c. etc. —— ——.

No. 17.

MR. P. B. BURNUM; Sur,

If you hav enny of them big Cokin Shiney fowl, that eat off tops of flour barils, I want sum. I gut a big nufoulan dogg that ways hard onto 140 pouns, and I want tu cell him, an git sum of them Cokin Shinys. This dogg is a gud dogg and dont eat much. I feed him on fish and hoggs hasslits and it dont cost much to keep him. He bitt a pedler's arm most off yisterday, but he woudnt be much trubble to you, if you kep him chaind *all the time* sose he couldnt bite nobody. If you will rite me what you ask for yure fowls, I will inform you what I ask for my dog. I dont want none nless thay can eat off tops of flour barrils,

* I would liked to have seen the dealer that could "fule" this customer more than "twict."

of course. Them 's the kind for me.* Anser by return mail. Yours Resp'y,

—— ——.

No. 18.

G. P. Burnham, Esq.:

I have got a Shanghae cock weighing 15½ pounds, and I want a few hens to match him. Can you supply me? My crower stands three feet four inches high, and his middle toe measures 7½ inches in length. What do you think of that? I want six twelve-pound hens. Dr. Bennett can supply me, I presume; but I want *pure*-bred stock. I have no doubt my crower will weigh eighteen or nineteen pounds, at two years old; he is now only eight months old! Let me hear from you.

Resp'y, —— — ——.

No. 19.

Mr. Burnham:

I always took you to be a man of honor, and I supposed *you* knew (if anybody did) what a Cochin-China fowl was, because you imported your stock. Now, those you sent me, and for which I willingly paid you $40 for the three, are feathered on the legs; this should *not* be, of course. How is this? They are fine, but I am certain they can be nothing but mere Shanghae fowls. Let me know about this, will you? Yours, &c.

—— — ——.

* I informed this purchaser that I could send him a pair which, if they "could n't eat off the tops" of his flour-barrels, I 'd warrant would eat up the *contents* of one as quickly as he could desire!

No. 20.

My Dear Sir :

I hardly know what to write you about the stock I had of you, six months ago, for "Cochin-Chinas." That they are *not* Cochins I feel positive, however; for one half the chickens came smooth-legged, and the rest are heavily-feathered on the legs ! ! I consider them only *Shanghaes*, and now I want to know if you can send me a trio of *pure* bloods, that you *know* to be Cochins. If so, I care nothing about price. I want *blood.* "Blood tells," you know. Let me hear from you, and state your own views in this matter. I will be governed by your advice. Enclosed is ten dollars for a dozen of your "Cochin" eggs —*pure*, you know. In haste,

Yours truly, —— ——.

No. 21.

Mr. Burnham.

Sir : Do you call yourself a man of honor ? I bought one doz. Cochin-China eggs of you, for which I paid you six dollars, cash. I set them, and I got but *ten* chickens out of them (two eggs I found rotten, in the nest). Every one of these chicks are cocks, sir — *cocks !* Now, what the devil can *I* do, do you imagine, with ten cocks ? I want to breed fowls. That is what I bought the eggs for ; to begin *right.* You must have *known* better than this. Anybody could have seen that these were all male eggs. *I* saw it, at once (I remember), but I *hoped* I was mistaken. What do you propose to do about this? Let me know, *at once*, without fail. In haste,

——— — ———.

No. 22.

SIR: You may think well of the Cochin-China fowls,—I *don't.* Those you sent me are long-legged, and there are no feathers on their legs, or feet, as there *ought* to be. *I* know what a Cochin-China fowl is, too well to be deceived in this way. I will keep them. *You are a humbug.* You are welcome to the thirty dollars I paid you. I don't ask you to return it. I don't want it. I can get along very well without it. You need it. Keep it. Much good may it do you! In haste,

——— ———.

P. S. Don't you wish you may get another $30 out of me, that way? O, yes — I guess you will — ha! ha!

No. 23.

MR. BARMAN. Dear Sir: I see in the Poultry Books that the Cotchin-China fowls lays two eggs every day,*

* "This gigantic bird," says Richardson, a noted English writer, "is very prolific, *frequently laying two, and occasionally three eggs on the same day!*" And, in support of this monstrous assertion, he subsequently refers, as his authority for this statement (which was called in

and sometimes three a-day. I have hens that lays two eggs a-day, frequenly, but I want to get the breed that will lay *three* eggs a-day, reglar. If you have got anny of the Cotchins that you *know* lays three eggs a-day, I would like to get a few, at a fair price. I don't pay no fancy prices for 'em, though. The hen fever won't larst forever, I don't believe; and then when its busted up, what 's the fowls good for, even if they *do* lay three eggs a-day? Let me hear from you,—but don't send any fowls unless you are *sure* they lay three eggs every day!

Yours, &c.,

—— ——.

No. 24.

MR. BURNHAM.—SIR: I am a gentleman, and I have no disposition to be fractious. I sent you twelve dollars, in a letter, for a dozen "Cotchin" eggs, and I set them. After waiting twenty-three days, I found two grizzled-colored chickens in the nest yesterday, both of them with huge *top-knots* on their polls! What does this mean? Am I to be swindled out of my money thus? By return of mail if you do not refund my money, if I live I will prosecute

question), to the "Rt. Hon. Mr. Shaw, Recorder of Dublin, to Mr. Walters, Her Majesty's poultry-keeper, and to J. Joseph Nolan, Esq., of Dublin." This was, in *my* opinion, one of the hums of the time, and I never had occasion to change that opinion. I do not believe the hen that *really* laid two eggs in one day ever lived to do it a second time! I have *heard* of this thing, however. But I never knew of the instance, myself.

you, if it costs me a thousand dollars. You may rely on this. I am not a man to be trifled with, and I refer you to Messrs. —— & ——, who know me; you evidently do *not!* In haste,

—— ——.

[I did not reply to this spicy favor, because, if the gentleman really was not a "fractious" man, I imagined he would like his pure-bred chickens better as they grew up; and, besides, I could afford to wait for "a gentleman" to cool off. I never heard from him, afterwards; and concluded that he did n't *live* to carry out his laudable intention of expending a thousand dollars in prosecuting me! I trust that, before he departed, he became hopefully pious. Peace to his manes!]

No. 25.

SIR: Them fouls you sent me, got the sore-hed. I gin em tuppentyn and unyuns and brandy, but it want no use. The poletry books sed so, and I follered the direction, and *it killed 'em both deader'n thunder, in one night!* Now you 've gut my mony, and I haint narry fowls. What 'll I do? Don't you think this a pooty impersition? Send me another pear, to once — if you don't want *fits.*

In haist,

—— ——.

[I sent this man "another *pear*,"— only I did n't!]

CHAPTER XIII.

THE OTHER SIDE OF THE QUESTION.

The foregoing are only a very few samples of hundreds upon hundreds of similar letters I constantly received, for nearly five years.

All the blame occasioned by careless express-men, of false blood imposed upon me originally, of tardy hens, of the hatching or non-hatching of eggs transported hundreds of miles, of feathered legs upon chickens, or the absence of them, of every species of mishap that could by any possibility befall the fancier and amateur, through his own ignorance or errors,—every kind of mistake was charged to *me!* But, with a Christian meekness, I bore it all.

I was threatened with civil prosecutions, with the House of Correction, the State Prison, the Penitentiary, and all sorts of other punishments, for my remissness; but I submitted with a quiet resignation, because "the people" were so deeply engaged in this pursuit, and everybody now had the fever so shockingly, that I sympathized with all mankind, and attributed these trifling ebullitions of ill-will, or raving, to the spasms caused by the prevalence and the severity of the epidemic.

On the other hand, I was so often cheered on in my labors of love by the kind consideration of a very different sort of patrons, that I did not sink under the persecution of those who would gladly have floored me, could the thing have been readily accomplished. I pocketed the money of my customers, however, bred good fowls, followed up the trade sharply, and found myself sailing easily along, in spite of the contemptible and small-fry opposition of which I was continually the object. As an agreeable offset to the complaints and murmurings in certain quarters, the following few letters will tell their own story:

From Hon. Henry Clay.

Ashland, 1851.

GEO. P. BURNHAM, ESQ., BOSTON.

MY DEAR SIR: I duly received your obliging letter, informing me that you had sent by the Express of Messrs. Adams & Co. a cage containing four fowls for me, and I postponed acknowledging it until the fate of the fowls should be ascertained. I have now the satisfaction to advise you that they all reached here safely.

They have been greatly admired, not only for their enormous size, but for their fine proportions and beautiful plumage. I thank you, my dear sir, most cordially, for this very acceptable present. It has been my aim, for many years, to collect at this place the best improved breeds of the horse, the cow, the sheep, swine and the ass — though the last, not the least valuable, in this mule-raising state.

To my stock on hand your splendid Cochin-China fowls will be a congenial and valuable addition; and, if we succeed with them, I will take care not to monopolize the benefit of them. I am greatly obliged to you, and,

With high respect, I am

Your obd't servant,

H. Clay.

From Gov. Geo. N. Briggs.

Pittsfield, 1851.

My Dear Sir:

The cage of Cochin-China chickens you were kind enough to send, reached me in safety; and I am much obliged to you for this favor.

They are, beyond comparison, the finest domestic fowls I have ever seen, and I shall breed them with such care that I hope to be able to give you a good account of them in the future.

They are very much liked by all who have seen them, and you will please accept my thanks for your attention.

I am, resp'y, yours,

Geo. N. Briggs.

From Hon. Daniel Webster.

Marshfield, 1851.

G. P. Burnham, Esq.

Dear Sir: The coop of chickens arrived safely, and are noble specimens of the Chinese fowl. You will rarely meet with samples apparently so well bred, and they will do any one credit. I thank you for the consignment, and consider them a most valuable addition to my stock of poultry. Accept my best wishes, and believe me, dear sir,

Yours, very truly,

Daniel Webster.

From Hon. Col. Phipps, H. B. M. Secretary.

Windsor Castle, Eng., 1853.

Dear Sir:

The cage of Grey Shanghae fowls intended as a present from you to Her Majesty the Queen has this day been received from Mr. Mitchell, of the Zoological Gardens, and they have been highly admired by Her Majesty.

I have received Her Majesty's commands to assure Mr. Burnham of her high appreciation of his attention; and to add that it affords another addition to the many marks of good will from the citizens of the United States which the Queen has received, and to which Her Majesty attaches so high a value. I have the honor to be

Your ob't and humble ser't,

C. B. Phipps.

Similar documents were often received by me, from friends and customers who knew how to appreciate good stock; and I have now hundreds of letters on file, of the most flattering character,— from *every* State in the Union, from England, Ireland, France, Bavaria, etc., where my stock was sent, and was roundly paid for,— all of which letters (with their enclosures, from time to time) served amply to "balance accounts" against the few received of an opposite character, and aided materially, also, to keep "the subscriber" from caving in!

Among the most friendly customers I ever had, and those who bought the most liberally,— while they were the most kindly in all their intercourse with me,— I must mention my patrons of the South generally, but especially the buyers in New Orleans and its vicinity. I never met with a trickster amongst them, and they paid me thousands upon thousands of dollars, without a word of cavil or complaint, from first to last. These fanciers had long purses, and are live *men*, with hearts "as big as a barn," so far as my experience goes.

CHAPTER XIV.

"BOTHER'EM POOTRUMS." BUBBLE NUMBER TWO.

THERE was something tangible, and *real*, in the "Cochin-China" fowl,—something that could be seen and realized (precious little, to be sure!), but still there was *something*. The Cochin-China hens would lay eggs (occasionally), and when they did n't breed their chickens with feathers upon the legs, they came without them. If the legs were not black or green skinned, they were either yellow or some *other* color. Their plumage was either spotted and speckled, or it was n't. And thus the true article, the *pure*-bred Cochins, could always be designated and identified,—by the knowing ones,—I *presume*. I studied them pretty carefully, however, for five years; but *I* never knew what a "Cochin-China" fowl really was, yet!

But when, in 1850 and '51, the "*Bother'ems*" begun to be brought into notice, I saw at once that, although this was bubble number two, it ought to have been number *one*, decidedly.

Never was a grosser hum promulgated than this was,

from beginning to end, even in the notorious hum of the hen-trade. There was absolutely nothing whatever in it, about it, or connected with it, that possessed the first shade of substance to recommend it, saving its *name*. And this could not have saved it, but from the fact that nobody (not even the originator of the unpronounceable cognomen himself) was ever able to write or spell it twice in the same manner.

The variety of fowl itself was the *Grey Chittagong*, to which allusion has already been made, and the *first* samples of which I obtained from "Asa Rugg" (Dr. Kerr), of Philadelphia, in 1850. Of this no one now entertains a doubt. They were the identical fowl, all over,—size, plumage and characteristics.

But my friend the Doctor wanted to put forth something that would take better than his "Plymouth Rocks;" and so he consulted me as to a name for a brace of *grey* fowls I saw in his yard. I always objected to the multiplying of titles; but he insisted, and finally entered them at our Fitchburg Dépôt Show as "*Burrampooters*," all the way from India.

These three fowls were bred from Asa Rugg's Grey Chittagong cock, with a yellow Shanghae hen, in Plymouth, Mass. They were an evident cross, all three of them having a *top-knot!* But, *n'importe*. They were then "Burrampooters."

Subsequently, these fowls came to be called "Burampootras," "Burram Putras," "Brama-pooters," "Brahmas," "Brama Puters," "Brama Poutras," and at last "Brahma Pootras." In the mean time, they were advertised to be exhibited at various fairs in different parts of the country under the above changes of title, varied in certain instances as follows: "Burma Porters," "Bahama Paduas," "Bohemia Prudas," "Bahama Pudras." And, for these three *last* named, prizes were actually offered at a Maryland fair, in 1851!

The following capital sketch (which appeared originally in the Boston *Carpet-Bag*) is from the pen of the late Secretary of the Mutual Admiration Society,—a gentleman, and a very happy writer in his way. It gives a faithful and accurate description of what many of these monsters really were, and will be read with gusto by all who have now come to be "posted up" in the secrets of the hen-trade.

The editor of the above-named journal remarks that "as our *Carpet-Bag* contains something connected with everything under the sun, we have abstracted therefrom a chapter on chicken-craft, which embraces a very important detail of that most abstruse science. When our readers scan the beautiful proportions of the stately fowl that *roosts* at the head of this article, they will acknowledge that we have some right to *cackle* because of the good fortune we have had in securing such an un*eggs*ceptionable picture, exhibiting the very perfection of cockadoodledom. Is n't he a beauty, this BOTHER'EM POOTRUM?

"Examine his altitude! Observe the bold courage that stands forth in his every lineament! There is no dunghill bravery there! See what symmetry floats round every detail of his noble proportions! What kingly grace associates with the comb that adorns his head as it were a crown! What fire there is in his eye! With what proud bearing does he not wear his abbreviated posterior appendage! Looking at the latter, we, and every one knowing in hen-craft, will readily exclaim, 'Gerenau de Montbeillard! you must have been a most unmitigated muff to designate *that* beautiful fowl the *gallus ecaudatus*, or tailless rooster.' For ourselves, our indignity teaches us to say, 'Mons. M.! your Essai sur Historie Nat. des Gallinacæ Fran. tom. ii., pp. 550 et 656, is a humbug!' We know that the universal world will sympathize in our sentiment on this point."

Peter Snooks, Esq. (a correspondent of this journal), it appears, had the honor to be the fortunate possessor of this invaluable variety of fancy poultry, in its unadulterated purity of blood. He furnished from his own yard samples of this rare and desirable stock for His Royal Highness Prince Albert, and also sent samples to several other noted potentates, whose taste was acknowledged to be unquestionable, including the King of Roratonga, the Rajah of Gabble-squash, His Majesty of the Cannibal Islands, and the Mosquito King. Peter supplies the annexed description of the superior properties of this variety of fowls:

"The *Bother'em Pootrums* are generally hatched from eggs. The original pair were not; they were sent from India, by way of Nantucket, in a whale-ship.

"They are a singularly *pictur-squee* fowl from the very shell. Imagine a crate-full of lean, plucked chickens, taking leg-bail for their liberty, and persevering around

Faneuil Hall at the rate of five miles an hour, and you have an idea of their extremely ornamental appearance.

"They are remarkable for producing bone, and as remarkable for producing offal. I have had one analyzed lately by a celebrated chemist, with the following result:

Feathers and offal, ...	39.00
Bony substances, ...	50.00
Very tough muscle and sinew, . .	09.00
Miscellaneous residuum, . . .	02.00
	100.00"

A peculiarly well-developed faculty in this extraordinary fine breed of domestic fowls is that of *eating.* "A tolerably well-fed Bother'em will dispose of as much corn as a common horse," insists Mr. S——. This goes beyond *me;* for I have found that they could be kept on the allowance, ordinarily, that I appropriated daily to the same number of good-sized store hogs. As to affording them *all* they would eat, I never did that. O, no! I am pretty well off, pecuniarily, but not rich enough to attempt any such fool-hardy experiment as that!

But Snooks is correct about one thing. They are not fastidious or "particular about *what* they eat." Whatever is portable to them is adapted to their taste for devouring. Old hats, India-rubbers, boots and shoes, or stray socks, are not out-of-the-way fare with them. They are amazingly fond of corn, especially *a good deal of it.* They *will* eat wheaten bread, rather than want.

They are very inquisitive in their nature. Their habit of stalking around the dwelling-house, and popping their heads into the garret-windows, is evidence of this peculiar trait.

Their flesh is firm and compact, and requires a great deal of eating to do it justice. Like Barney Bradley's leather "O-no-we-never-mention-'ems," when cut up and stewed for tripe, "a fellow could eat a whole bushel of potatoes to the plateful." It is of the color of a stale red herring, and very much like that edible in taste. Its scarcity constitutes its value.

This *rara avis in terris* grows to a height somewhere between .00 feet .16 inches and 25 feet. Its weight somewhat between .06 pounds and 1 cwt. It never lays, except when it rolls itself in the sand. The female fowls sometimes do that duty, though amazingly seldom.

Mr. Snooks says he will back his Bother'em, for a chicken-feast, to outcrow any three asthmatical steam-whistles that any railroad company can scare up; and adds, "I am ashamed of the prejudice which makes my fellow-men unjust. The Fowl Society — the New England organization, I mean — repudiate the special merits of my *Bother'em*

Pootrums, and tell me that their ideas of improvement go entirely contrary to the propriety of tolerating my noble breed of fowls. *Disgustibus non disputandum*, as Shakspeare, or somebody for him, emphatically says,—which means, 'Every one to his taste, as the old lady said when she kissed the cow.' One thing it will not be hard to prove, I think; that is, simply the probability of something like envy operating among the members of the Hen Society, on account of the exclusive attention paid my *Bother'ems* at the late Fowl Fairs in Boston,"—where the 'squire's contributions *did* rather "astonish the boys" who were not thoroughly acquainted with the excellent qualities of these birds. Verily, Snooks' "Bother'ems" did bother 'em exceedingly!

CHAPTER XV.

ADVERTISING EXTRAORDINARY.

FROM the outset of my experience in the final attack of the hen fever, I took advantage of every possible opportunity to disseminate the now world-wide known fact that nobody else but myself possessed any "pure-bred" poultry! I could have proved this by the affidavits of more than a thousand "disinterested witnesses," at any time after April and May, 1851, had I been called upon so to do. But as no one *doubted* this, there was then no controversy.

But, as time wore along, competition became rife, and the foremost chicken-raisers began to look about them for the readiest means obtainable with which to cut each other's throats; not "with a feather," by any means, because that would have "smelt of the shop;" but whenever, wherever, or however, their neighbors could be traduced, maligned, vilified, or injured (in this pursuit), they embraced the opportunity, and followed it up, without stint, especially towards my humble self, until most of them, fortunately, broke their own backs, and were compelled to retire from the field, while " the people " grinned, and comforted them with the friendly assurance that it " sarved 'em right."

At the Fitchburg Dépôt Show, in 1850, my original " Grey Chittagongs " (already described) were in the possession of G. W. George, Esq., of Haverhill, to whom they had been sold by the party to whom I had previously sold them. Nobody thought well of them; but they took a first prize there, and the " Chittagongs " (so entered at the same time) of Mr. Hatch, of Connecticut, also took a prize. My friend the Doctor then insisted that these were *also* " Burrampooters; " but, as nobody but himself could pronounce this jaw-cracking name, it was taken little notice of at that time.

Mr. Hatch had a large quantity of the Greys at this show, which sold readily at $12 to $20 the pair; and immediately after this exhibition the demand for " Grey Chittagongs " was very active. I watched the current of the stream, and I beheld with earnest sympathy the now alarming symptoms of the fever. " The people " had suffered a relapse in the disease, and the ravages now promised to become frightful — for a time!

An ambitious sea-captain arrived at New York from Shanghae, bringing with him about a hundred China fowls, of all colors, grades, and proportions. Out of this lot I selected a few *grey* birds, that were very large, and (consequently) "very fine," of course. I bred these, with other grey stock I had, at once, and soon had a fine lot of birds to dispose of — to which I gave what I have always deemed their only true and appropriate title (as they came from Shanghae), to wit, *Grey Shanghaes.*

In 1851 and '52 I had a most excellent "run of luck" with these birds. I distributed them all over the country, and obtained very fair prices for them; and, finally, the idea occurred to me that a present of a few of the choicest of these birds to the Queen of England would n't prove a very bad advertisement for me in this line. I had already reaped the full benefit accruing from this sort of "disinterested generosity" on my part, toward certain *American* notables (whose letters have already been read in these pages), and I put my newly-conceived plan into execution forthwith.

I then had on hand a fine lot of fowls, bred from my "imported" stock, which had been so much admired, and I selected from my best "Grey Shanghae" chickens nine beautiful birds. They were placed in a very handsome black-walnut-framed cage, and after having been duly lauded by several first-rate notices in the Boston and New York papers, they were duly shipped, through Edwards, Sanford & Co.'s Transatlantic Express, across the big pond, addressed in purple and gold as follows:

TO H. M. G. MAJESTY,

Victoria,

QUEEN OF GREAT BRITAIN.

To be Delivered at Zoological Gardens,

LONDON, ENG.

FROM GEO. P. BURNHAM, BOSTON, MASS., U. S. A.

The fowls left me in December, 1852. The *London Illustrated News* of January 22d, 1853, contained the following article in reference to this consignment:

"By the last steamer from the United States, a cage of very choice domestic fowls was brought to Her Majesty Queen Victoria, a present from George P. Burnham, Esq., of Boston, Mass. The consignment embraced nine beautiful birds — two males and seven pullets, bred from stock imported by Mr. Burnham direct from China. The fowls are seven and eight months old, but are of mammoth proportions and exquisite plumage — light silvery-grey bodies, approaching white, delicately traced and pencilled with black upon the neck-hackles and tips of the wings and tails. The parent stock of these extraordinary fowls weigh at maturity upwards of twenty-three pounds per pair; while their form, notwithstanding this great weight, is unexceptionable. They possess all the rotundity and beauty of the Dorking fowl; and, at the same age, nearly double the weight of the latter. They are denominated Grey Shanghaes (in contradistinction to the Red or Yellow Shanghaes), and are considered in America the finest of all the great Chinese varieties. *That they are a distinct race, is evident from the accuracy with which they breed, and the very close similarity that is shown amongst them; the whole of these birds being almost precisely alike, in form, plumage and general characteristics.* They are said to be the most prolific of all the Chinese fowls. At the time of their shipment, these birds weighed about twenty pounds the pair."

This was a very good *beginning.* In another place (see page 88) I have given a copy of the letter from Hon. Col. Phipps, her Majesty's Secretary of the Privy Purse, acknowledging the receipt of this present. A few weeks afterward, the *London News* contained a spirited original picture of seven of the nine Grey Shanghae fowls which I had the honor to forward to Queen Victoria. The drawing was made by permission of the Queen, at the royal poultry-house, from life, by the celebrated *Weir*, and the engraving was admirably executed by *Smythe*, of London. The effect in the picture was capital, and the likenesses very truthful. In reference to these birds, the *News* has the following:

"GREY SHANGHAE FOWLS FOR HER MAJESTY.—In the *London Illustrated News* for January 22d, we described a cage of very choice domestic fowls, bred from stock imported by Mr. George P. Burnham, of Boston, Mass., direct from China, and presented by him to Her Majesty. We now engrave, by permission, these beautiful birds. They very closely resemble the breed of *Cochin-Chinas* already introduced into this country, the head and neck being the same; the legs are yellow and feathered; the carriage very similar, but the tail being more upright than in the generality of Cochins. The color is creamy white, slightly splashed with light straw-color, with the exception of the tail, which is black, and the hackles, which are pencilled with black. The egg is the same color and form as that of the Cochins hitherto naturalized in this country. These fowls are very good layers, and have been supplying the royal table since their reception at the poultry-house, at Windsor."

All this "helped the cause along" amazingly. It proved a most excellent mode of advertising my "superb," "magnificent," "splendid," "unsurpassable," "inapproachable" GREY SHANGHAES.

The above articles found their way (somehow or other) into the papers of this country immediately; and, within sixty days afterwards, the price of "Bother'ems" went up from $12 and $15 to $50, $75, $100, and $150, the pair!!

"Cochin-Chinas" were now *no*whar! But *I* was so as to be about yet.

CHAPTER XVI.

HEIGHT OF THE FEVER.

WHILE this cage of Grey Shanghaes stood for an hour or two in the express-office of Adams & Co., in Boston, a servant came from the Revere House to inform me that "a gentleman desired to see me there, about some poultry."

As I never had had occasion to run round much after my customers, and, moreover, as I felt that the dignity of the business — (the *dignity* of the hen-trade !) — might possibly be compromised by my responding in person to this summons, I directed the servant to "say to the gentleman, if he wished to see me, that I should be at my office, No. 26 Washington-street, for a couple of hours,— after that, at my residence in Melrose."

The man retired, and half an hour afterwards a carriage stopped before my office-door. The gentleman was inside. He invited me to ride with him — (I could afford to *ride* with him) — to Adams & Co.'s office. He had seen the "Grey Shanghaes" intended for the Queen there.

"I want that cage of fowls," he said.

"My dear sir," I replied, "they are going to England."

"I *want* them. What will you take for them?"

"I can't sell them, sir."

"You can send others, you know."

"No, sir. I can't dispose of *these*, surely."

"Can you duplicate this lot?"

"Pretty nearly — perhaps not quite."

"I see," he continued. "I will give you two hundred dollars for them."

"No, sir."

"Three hundred — come!"

"I can't sell them."

"Will you take *four* hundred dollars for the nine chickens, sir?" he asked, drawing his pocket-book in presence of a dozen witnesses.

I declined, of course. I could n't sell these identical fowls; for I had an object in view, in sending them abroad, which appeared to me of more consequence than the amount offered — a good deal.

"Will you *name* a price for them?" insisted the stranger.

I said, "No, sir — excuse me. I would not take a thousand dollars for these birds, I assure you. Their equals in quality and number do not live, I think, to-day, in America!"

"I won't give a — a — thousand dollars, for them," he

said, slowly. "No, I won't give *that!*" and we parted. Yet, I have no doubt, had I encouraged him with a prospect of his obtaining them at all, he *would* have given me a thousand dollars for that very cage of fowls! To *this* extent did the hen fever rage at that moment.

I subsequently sent this gentleman two trios of my grey chickens, for which he paid me $200.

And now the Grey Shanghae trade commenced in *earnest.* Immediately after the announcements were made (which I have quoted) orders poured in upon me furiously from all quarters of this country, and from Great Britain. Not a steamer left America for England, for months and months, on board of which I did not send more or less of the "Grey Shanghaes." From every State in the Union, my orders were large and numerous; and letters like the following were received by me almost every day, for months:

"G. P. BURNHAM.

"SIR: I have just seen the pair of superb Grey Shanghae fowls which you sent to Mr. —— ——, of this city, and I want a pair like them. If you can send me *better* ones, I am willing to pay higher for them. He informs me that your price per pair is forty dollars. I enclose you *fifty* dollars; do the best you can for me, but forward them *at once,*— don't delay. Yours, &c.,

"—— —— ——."

I almost always had "*better* ones." That was the kind I always kept behind, or for my own use. I rarely sent away these better ones until they cried for 'em! I always had a great *many* of the "best" ones, too; which were even better than those "better" ones for which the demand had come to be so great!

Strange to say, everybody got to want *better* ones, at last; and, finally, I had none upon my premises but this very class of birds — to wit, the "better ones." To be sure, I reserved a very *few* pairs of the *best* ones, which could be obtained at a fair price; but these were the ones that would "take down" the fanciers, occasionally, who wanted to beat *me* with them at the first show that came off. But I did n't sleep much over this business. I always had one cock and three or four hens that the boys did n't *see* — until we got upon the show-ground. Ha, ha!

A stranger called at my house, one Sunday morning, just as I was ready with my family for church. He apologized for coming on that day, but could n't get away during the week. He had never seen the Grey Shanghaes — did n't know what a Chinese fowl was — had no idea about them at all. He wanted a few eggs — heard I had them — would n't stop but a moment — saw that I was just going out, &c. &c. He sat down — was sorry to trouble me — would n't do so again — would like just to take a peep at the fowls — when, suddenly, as he

"I DON'T WANT ANY EGGS—NO!"—(See page 109.)

sat with his back close to the open window, my old crower sent forth one of those thundering, unearthly, rolling, guttural shrieks, that, once heard, can never be forgotten!

The stranger leaped from his chair, and sprang over his hat, as he yelled,

"Good God! what 's *that?*"

His face was as white as his shirt-bosom.

"That 's one of the Grey Shanghaes, crowing," I replied.

"*Crow!* I beg your pardon," he said; "I don't want any eggs — no! I 'll leave it to another time. I — a — I could n't take 'em now; won't detain you — good-morning, sir," he continued; and, rushing out of my front door, he disappeared on "a dead run," as fast as his legs could carry him. And I don't know but he is running yet. He was desperately alarmed, surely!

I was so amused at this incident, that I was in a precious poor mood to attend church that morning. And when my friend the minister arose at length, and announced for his text that "the wicked flee when no man pursueth," those words capped the climax for me.

I jammed my handkerchief into my mouth, until I was nearly suffocated, as I thought of that wicked fellow who had just been so frightened while in the act of attempting to bargain for fancy hen's eggs on the Sabbath!

A Western paper, in alluding to the fever, about this period, observed that "this modern epidemic has shown itself in our vicinity within a short time, and is characterized by all the peculiarities which have marked its ravages elsewhere. Some of our most valuable citizens are now suffering from its attacks, and there is no little anxiety felt for their recovery. The morning slumbers of our neighbors are interrupted by the sonorous and deep-toned notes of our Shanghae Chanticleer, and various have been the inquiries as to how he took '*cold*,' and what we gave him for it. 'Chittagongs' and 'Burma Porters' are now as learnedly discussed as 'Fancy Stocks' on change.

The N. Y. *Scientific American* stated, at this time, that the "Cochin-China fowl fever was then as strong in England as in some parts of New England,—in fact, stronger. One pair exhibited there was valued at $700. What a sum for a hen and rooster! The common price of a pair is $100," added this journal; and still the trade continued excellent with me.

CHAPTER XVII.

RUNNING IT INTO THE GROUND.

THERE now seemed to be no limit whatever to the *prices* that fanciers would pay for what were deemed the best samples of fowls. For my own part, from the very commencement I had been considerate and merciful in *my* charges. True, I had been taken down handsomely by a Briton (in my original purchase of Cochin-Chinas), but I did not retaliate. I was content with a fair remuneration; *my* object, principally, was to disseminate good stock among "the people," for I was a democrat, and loved the dear people.

So I charged lightly for my "magnificent" samples, while other persons were selling second and third rate stock for five or even six and eight dollars a pair. The "Grey Shanghaes" had got to be a "fixed fact" in England, as well as in this country, and still I was flooded with orders continually.

I obtained $25, $50, $100 a pair, for mine; and one gentleman, who ordered four greys, soon after the Queen's stock reached England, paid me *sixty guineas* for them — $150 a pair. But these were of the *better* class of birds to which I have alluded.

In 1852 a Boston agricultural journal stated that "within three months extra samples of two-year-old fowls, of the large Chinese varieties, have been sold in Massachusetts at $100 the pair. Several pairs, within our own knowledge, have commanded $50 a pair, within the past six months. Last week we saw a trio of White Shanghaes sold in Boston for $45. And the best specimens of Shanghaes and Cochin-China fowls now bring $20 to $25 a pair, readily, to purchasers at the South and West."

Now, these prices may be looked upon by the uninitiated as extraordinary. So they were for this country. But at a Birmingham (Eng.) show, in the fall of 1852, a single pair of "Seabright Bantams," very small and finely plumed, sold for $125; a fine "Cochin-China" cock and two hens, for $75; and a brace of "White Dorkings," at $40. An English breeder went to London, from over a hundred miles distant, for the sole purpose of procuring a setting of Black Spanish eggs, and paid one dollar for each egg. Another farmer there sent a long distance for the best Cochin-China eggs, and paid one dollar and fifty cents *each* for them, at this time!

This was keeping up the rates with a vengeance, and beat us Yankees, out and out. But later accounts from across the water showed that this was only a beginning, even. In the winter of 1852 the *Cottage Gardener* stated that "within the last few weeks a gentleman near London sold a pair of Cochin-China fowls for 30 guineas ($150), and another pair for 32 guineas ($160). He has been offered £20 for a single hen; has sold numerous eggs at 1 guinea ($5) each, and has been paid down for chickens just hatched 12 guineas ($60) the half-dozen, to be delivered at a month old. One amateur alone had paid upwards of £100 for stock birds."

To this paragraph in the *Gardener* the *Bury and Norwich Post* added the following: "In our own neighborhood, during the past week, we happen to know that a cock and two hens (Cochin-Chinas) have been sold for 32 guineas, or $160. The fact is, choice birds, well bred, of good size and handsome plumage, are now bringing very high prices, everywhere; and the demand (in our own experience) has never been so great as at the present time."

In this way the fever raved and raged for a long year or more. Shows were being held all over this country, as well as in every principal city and town in England. Everybody bought fowls, and everybody had to pay for them, too, in 1852 and 1853!

In a notice of one of the English shows in that year (1853), a paper says: "There is a pen of three geese weighing forty-eight pounds; and among the *Cochin-China* birds are to be found hens which, in the period that forms the usual boundary of chicken life, have attained a weight of seven or eight pounds. Of the value of these birds it is difficult to speak without calling forth expressions of incredulity. It is evident that there is a desperate *mania* in bird-fancying, as in other things. Thus, for example, there is a single fowl to which is affixed the enormous money value of 30 *guineas;* two Cochin-China birds are estimated at 25 *guineas;* and four other birds, of the same breed, a cock and three hens, are rated in the aggregate at 60 *guineas,*— a price which the owner confidently expects them to realize at the auction-sale on Thursday. A further illustration of this ornithological enthusiasm is to be found in the fact that, at a sale on Wednesday last, one hundred and two lots, comprising one hundred and ten Cochin-China birds, all belonging to one lady, realized £369. 4s. 6d.; the highest price realized for a single one being 20 guineas."

Another British journal stated, a short time previously, that "a circumstance occurred which proves that the Cochin-China mania has by no means diminished in intensity. The last annual sale of the stock of Mr. Sturgeon, of Greys, has taken place at the Baker-street Bazaar. The two hundred birds there disposed of could not have realized a less sum than nearly £700 (or $3500), some of the single specimens being knocked down at more than £12, and very many producing £4, £5, and £6 each."

The attention, at this sale, devoted to the pedigree of the birds, was amusing to a mere observer; one fowl would be described as a cockerel by *Patriarch*, another as a pullet by *Jerry*, whilst a third was recommended as being the offspring of *Sam*. Had the sale been one of horses, more care could hardly have been taken in describing their pedigrees or their qualifications. Many were praised by the auctioneer as being particularly *clever* birds, although in what their cleverness consisted did not appear. The fancy had evidently extended to *all* ranks in society. The peerage sent its representatives, who bought what they wanted, regardless of price. Nor was the lower house without its delegates; a well-known metropolitan ex-member seems to have changed his constituency of voters for one of Cochins; and we can only hope that it may not be his duty to hold an inquest on any that perish by a violent or unnatural death. The sums obtained for these birds depended on their being in strict accordance with the then taste of the fancy. They were magnificent in size, docile in behavior, intelligent in expression, and most of them were very finely bred.

And while the hen fever was thus at its height, almost, in England, we were following close upon the footsteps of John Bull in the United States. At the Boston Fowl

Show in 1852, three Cochin-Chinas were sold at $100; a pair of Grey Chittagongs, at $50; two Canton Chinese fowls, at $80; three Grey Shanghae chicks, at $75; three White Shanghaes, at $65; six White Shanghae chickens, $40 to $45, etc.; and these prices, for similar samples, could have been obtained again and again.

At this time there was found an ambitious individual, occasionally, who got "ahead of his time," and whose laudable efforts to oustrip his neighbors were only checked by the natural results of his own superior "progressive" notions. A case in point:

"Way down in Lou'siana," for instance, a correspondent of mine stated that there lived one of these go-ahead fellows, who had been afflicted with a serious attack of hen fever, and who was not content with the ordinary speed and prolificness in breeding of the noted Shanghae fowls. He desired to possess himself of the biggest kind of a pile of chickens for the rapidly augmenting trade; and so he had constructed an Incubator, of moderate dimensions, into which he carefully stowed only three hundred nice fresh eggs, from his fancy fowls.

The secret of his plan to "astonish the boys" was limited to the knowledge of only two or three friends; and — thermometer in hand — he commenced operations. With close assiduity and Job-like patience, our amateur applied himself to his three weeks' task, by day and night, and at the end of fifteen days, one egg was broken, and Mr. Shanghae was *thar*,— alive and kicking, but as yet immature.

The neighborhood was in the greatest excitement at this prospect of success. Our friend commenced to crow (slightly), and, to hasten matters, put on a *leetle* more steam at a venture. The twenty-second day arrived, and the "boys" assembled to witness the *entrée* of three hundred steam-hatched Shanghaes into this breathing world. Our amateur was full of expectation and "fever." One egg was broken; another, and then another; when, upon inspection, the entire mass was found to have been *thoroughly boiled!*

A desperate guffaw was heard as our amateur friend disappeared, and his only query since has been to ascertain what actual time is required to boil a certain quantity of eggs at a given heat, and the smallest probable cost thereof! As far as heard from, the reply has been, say six gallons of good alcohol, at one dollar per gallon, for three hundred eggs; time (night and day), twenty-two days and seven hours; and the product it is generally thought would make capital fodder for young turkeys,— provided said eggs are not boiled *too hard!*

On the subject of the *diseases* of poultry many learned and sapient dissertations appeared about these days. In one agricultural journal we remember to have met with the following scientific prescription. The learned writer is talking about *roup* in fowls, and says:

"This is probably a chronic condition, the result of frequent colds. Give the following medicines: *Aconite*, if there is fever, *hepar-suliphuris third trituration*, or mercury, *third trituration*, for a day or two, once in three or four hours; then *pulsatilla tincture* for the eyes; *antimonium*, third trituration or *arsenic*, or *nux vomica*, for the crop."

Is n't this *clear*, reader? How many poultry-raisers in the United States are there who would be likely to comprehend one line of this stuff? We advise this writer to try again; the above is an "elegant extract," verily!

We now come down to the fourth and last exhibition in Boston of the Mutual Admiration Society, *alias* the Association with the long-winded cognomen, which took place in September, 1852.

CHAPTER XVIII.

ONE OF THE FINAL KICKS.

I WAS chosen by somebody (who will here permit me to present them my thanks for the honor) as one of the judges to decide upon the merits of the birds then to be exhibited: and my colleagues on this Committee were Dr. J. C. Bennett, and Messrs. Andrews, Balch and Fussell.

On the morning of the opening of this show the names of the judges were first announced to the contributors. Immediately there followed a "hullabaloo" that would have done credit to any bedlam, ancient or modern, ever heard or dreamed of. The lead in this burst of rebellion amongst the hitherto "faithful" was taken by one prominent member, who announced publicly, then and there, that the selection of the judges was an infamous imposition. They were incompetent, dishonest, prejudiced, calculating, speculative, ambitious competitors. Moreover, that it had all been "contrived by that damned Burnham, who would rob a church-yard, or steal the cents off the eyes of his dead uncle, any time, for the price of a hen."

These were the gentleman's own expressive words. He added that he could stand anything in the hen-trade but *this*. This, however, he would *not* submit to. Burnham should be kicked out of that Committee, or he would kick himself out of his boots, and the Society's traces also; — a threat which did not seem to alarm or disturb anybody, "as I knows on," except this same tall, stout, athletic, brave, honorable, honest, truthful, smart, gentlemanly member of this Mutual Admiration Society!

Now, it was very well known, at this time, that the Committee of Judges had been chosen entirely without their own knowledge. So far as I was myself concerned, I should greatly have preferred at that time to have remained an outsider, because it would have then been quite as well for me to have contributed to the exhibition, where, with the "splendid specimens" I then possessed of the Cochin-China and Shanghae varieties of fowl, I could have knocked all the others "higher than a fence" in *that* show, as I had done in all the previous exhibitions where I had ever competed with the boys.

But the same power which had formed the Committee of Judges also provided that they must not be competitors. Thus, three or four of those persons who had at the previous exhibitions of this Society been the most extensive contributors,— men who had bred by far the largest assortments and quantities of good fowls up to this period, and who had till now paid ten or twenty dollars for one (compared with any other of the members) toward the good of the association, and in the furtherance of its objects,—*these* men were made the judges, and were cut off as contributors. I was satisfied, however, because I saw that the framing of the *Report* of this show would fall to my lot again; and I had no doubt that, under these circumstances, I could afford to be "persecuted" for the time being.

It is not in my nature to harm anybody; and those who are personally acquainted with me, know that I am *constitutionally* of a calm, retiring, meek, religious turn of mind. My aim in life is to "do unto others as I would have others do unto me." I "love my neighbor" (if he does n't permit his hens to get into my garden) "as myself." And, "if a man smite me upon one cheek, I turn to him the other also," immediately, if not sooner. I never retaliate upon an enemy or an opponent —until I make *sure that I have him where the hair is short.*

I once knew of an extraordinary instance of patience that taught me a powerful lesson in submissiveness. It occurred in a Western court, where the judge (a most exemplary man, I remember) sat for two mortal days quietly listening to the arguments of a couple of contending lawyers in reference to the construction they desired him to assume in regard to a certain act of the Legislature of that State. When the two legal gentlemen had "thrown themselves," in this long and wearying debate, for forty-eight hours, his Honor cut off the controversy by remarking, very quietly,

"Gentlemen, this law that you have been speaking of *has been repealed!*"

I thought of this circumstance, and I permitted the hen-men to gas, to their hearts' content. When they got through with their anathemas, their spleen, and their stupidity, I informed them that the "Committee" had unanimously left to *my* charge *the writing of the Report of that Exhibition.*

From that moment, up to the hour when the Report was published, I never suspected (before) that I had so many *friends* in this world!

The fear that seemed to pervade every mind present was, that *I* should probably do precisely what *they* would have done under similar circumstances,— to wit, take care of myself.

I had no fowls in this exhibition; but there were present numerous specimens bred from my stock, that were very choice (so every one said), and which commanded the highest prices during the show.

There were several *Southern* gentlemen present, who bought (and paid roundly for them, too) some of the best fancy-birds on sale. It was astonishing how much some of those buyers did know about the different breeds of Chinese fowls there! Yes, it certainly was astounding! I think I *never* saw before so much real, downright *bona fide* knowledge of henology displayed as was shown by one or two Southern gentlemen, then and there; — never, in the whole course of my experience!

By reference to the next chapter, it will be seen how shamefully I neglected my own interests, and how self-sacrificing I was in the report of the Society's last kick, which, as I have already hinted, the Committee left to *my* charge to prepare.

I had no disposition (in the preparation of this document) to underrate the stock of any one else, *provided* it did not interfere with me! And, after carefully noting down whatever seemed of importance to my well-being there, I sat myself down to oblige the Committee by writing the "Report" of this show, which an ill-natured competitor subsequently declared was "only in favor of Burnham and his stock, all over, underneath, in the middle, outside, overhead, on top, on all sides, and at both ends!"

And *I believe he was right!*

CHAPTER XIX.

THE FOURTH FOWL-SHOW IN BOSTON.

THIS show (in September, 1852) was the fifth exhibition held in Boston, but the *fourth* only of the Society with the long name.

The Report commences with a congratulation (as usual) that the association still lives, and has a being; and, after alluding to the general state of the affairs of the concern,— without touching upon its financial condition,— it thus proceeds:

"Your Committee would call your attention to the fact that among the numerous fowls exhibited this season,— as upon former occasions,— a very unnecessary practice seems to have obtained, in the mis-*naming* of varieties. Cross-bred fowls have been called by original cognomens, unknown to practical breeders; and a host of birds well known to the Committee, as well as to poulterers generally, have been denominated by any other than their *real* and universally conceded ornithological titles. This savors of bad taste; it leads to ridicule among strangers who visit our shows from abroad; and should not be sanctioned by your Society. Errors may creep in among your transactions, in this particular, and many honest, careful breeders may be deceived; but the multiplying of *unpronounceable and meaningless names* for domestic fowls is entirely uncalled for; and your committee recommend a close adherence, hereafter, to recognized titles only.

"In this connection, it may be proper to allude to a case in point. The largest and unquestionably one of the finest varieties of domestic fowls ever shown among us was entered by the breeders of this variety as the 'Chittagong;' other coops of the same stock were labelled 'Grey Chittagongs;' others were called 'Bramah Pootras;' and others, 'Grey Shanghae' and 'Malays.'

"Your Committee are divided in opinion as to what these birds ought, rightfully, to be called,— though the majority of the Committee have no idea that 'Bramah Pootra' is their correct title. That they are not 'Malays' is also quite as clear. Several of the specimens are positively known to have come direct from Shanghae; and *none* are known to have come originally from anywhere else. Nevertheless, it has been thought proper to leave this question open, for the present; and the Committee, believing that this fowl originates in and hails directly from the East, are content to accept for them the title of 'Grey Shanghae,' 'Chittagong,' or 'Bramah Pootra,' as different breeders may elect,— admitting, at the same time, that they are really a very superior bird, and believing that if carefully bred they may be found decidedly the most valuable among all the large *Chinese* breeds, of which they are clearly a good variety."

* * * * * *

"A large sum of money was expended at this exhibition, by visitors, amateurs and breeders,— one gentleman investing upwards of $700 in choice fowls; another, from the South, purchasing to the amount of $350 for extra samples; another bought $200 worth, etc. The highest figures ever yet paid on this side of the Atlantic (for individual purchases) were realized at this show.

"Samples of the China stock originally imported from Shanghae were very plentiful on this occasion, and the high reputation of this blood was fully sustained in the specimens exhibited. Very superior fowls, bred from G. P. Burnham's importations of Cochin-Chinas, were also numerous, and were sold, in four or five instances, at the very *highest* prices paid for any samples that were disposed of."

Among the premiums awarded to the *Chinese* fowls by this "Committee," were the following:

"CHINA FOWLS. — To H. H. Williams, best cock and two hens (of Burnham's *Canton* importation), $5. To C. Sampson, West Roxbury, best cock and single hen (Burnham's *Canton* importation), $3. To H. H. Williams, third prize, for same stock, $2. To C. C. Plaisted, Great Falls, N. H., the Committee awarded a first prize, $5, for what he called '*Hong-Kong*' fowls; these were of Burnham's *Canton* stock, also. To A. White, E. Randolph, for six best chickens (Burnham's importation), $2.

"COCHIN-CHINA. — To H. H. Williams, West Roxbury, best cock and two hens (splendid samples, of extraordinary size and beauty), first prize, $5. To A. White, E. Randolph, best cock and single hen (of Burnham's importation), $3. To A. White, for six best chickens (Burnham's importation), $2."

* * * * * *

The Committee then allude to the *prices* which were paid there for fowls, "*not because they advocate the propriety of keeping them up*" (O, no!), "but rather to show that the welfare of the Association is by no means derogating.

"The three prize *Cochin-China* fowls were sold for $100. The two prize *Grey Shanghaes*, or 'Bramah Pootras,' were sold for $50. Three chickens of the same, at $50. A pair of Burnham's importation of *Cochins*, at $80; another pair, at $40; another trio (chickens), at $40. Six Black Spanish chickens (Child's), at $50. Six *White Shanghae* chickens (Wight's), at $45. Three hens, of same stock, at $50 — and several pairs and trios of other varieties, at from $25 each, to $25 and $30 to $40 the lot."

* * * * * *

At a subsequent meeting of the Trustees, Mr. George P. Burnham, on the part of the Judges at the late exhibition of the Society, presented their *Report*, whereupon it was

"*Voted*, That the Report of the Judges on the recent show of poultry in the Public Garden be accepted."

And this was the end of *that* ball of worsted! I rather have the impression, now,—as nearly as I can recollect (though my memory is somewhat treacherous in these matters), but I *think* I sold a few fowls, just after that fair. "I may be mistaken,—but that is my opinion!"

The Report was duly accepted, in form, and I had the satisfaction of seeing my "extraordinary" and "superb" stock again lauded to the very echo, at the expense of the old-fogyism of the "Mutual Admiration Society."

The consequence was a renewed activity in my sales, which continued delightfully lively and correspondingly remunerative for several months after *this* exhibition, also, where I did not enter the first fowl!

CHAPTER XX.

PRESENT TO QUEEN VICTORIA.

I HAVE already alluded to the fine Grey Shanghaes which I forwarded to Her Majesty the Queen. In relation to this circumstance the Boston papers contained the following announcement, in the month of April, 1853; a circumstance which did not greatly retard the prospects of my business either on this or on the other side of the water! The compliment thus paid me by Royalty was duly appreciated, and its delicacy will be apparent to the reader. This picture is the only one of its *kind* ever sent to an American citizen.

"A COMPLIMENT FROM VICTORIA. — Some weeks ago, Mr. George P. Burnham, of Boston, forwarded to Her Majesty Queen Victoria a present of some *Grey Shanghae* fowls, which have been greatly admired in England. By the last steamer Mr. Burnham received the following letter from Her Majesty's Secretary of the Privy Purse, accompanying a fine portrait of the Queen, sent over to Mr. B.:

The Queen's Letter.

"Buckingham Palace,
March 15, 1853.

"Dear Sir: I have received the commands of Her Majesty the Queen, to assure you of Her Majesty's high appreciation for the kind motives which prompted you to forward for her acceptance the magnificent 'Grey Shanghae' fowls which have been so much admired at Her Majesty's aviary at Windsor.

"Her Majesty has accepted, with great pleasure, such a mark of respect and regard, from a citizen of the United States.

"I have, by Her Majesty's command, shipped in the 'George Carl,' to your address, a case containing a portrait of Her Majesty,* of which the Queen has directed me to request your acceptance.

"I have the honor to be,

"Sir, your ob't and humble servant,

"C. B. Phipps.

"To Geo. P. Burnham, Esq.,
Boston, U. S. A."

* See Frontispiece.

I caused a copy to be taken from this portrait of the Queen, and have had it engraved for this book; it appears as the frontispiece.

Immediately after this paragraph appeared, a new zest appeared to have been given to the Grey Shanghae trade. Orders came from Canada and from Nova Scotia to a very considerable amount; and during this season my sales were again very large. During the year 1853, I started and raised over sixteen hundred chickens of all kinds; but this did not supply my orders. I bought largely, and paid high prices, too, generally. But few persons were now doing any business in the fowl-trade, except myself, however.

The *N. Y. Spirit of the Times* published portraits of the birds sent to the Queen, and remarked that "the engraving represented six of the nine beautiful *Grey Shanghae* fowls lately presented to Her Majesty *Victoria*, Queen of Great Britain, by *George P. Burnham*, Esq., of Boston, Mass.

"These birds were forwarded by one of the last month's Collins steamers, in charge of Adams & Co.'s Express, and passed through this city on the 24th ult. Their extraordinary size and fine plumage were the admiration of all who examined them. The picture is from life, engraved by Brown, and is a faithful representation of the birds, which are very closely bred.

"The color of this variety of the China fowl is a light

silver-grey, approximating to white; the body is a light downy white, sparsely spotted and pencilled with metallic black in the tail and wing tips; the legs are feathered to the toes, and the form is unexceptionable for a large fowl; this variety having proved the biggest of all the 'Shanghaes' yet imported into this State.

"The two cocks above delineated weighed between ten and eleven pounds each at six months old; the pullets drew seven and a half to nine pounds each at seven to eight months old; the original imported pair of *old* ones now weigh upwards of twenty-three pounds, together. In the existing rage for weighty birds, this variety will naturally satisfy the ambition of those who go for the 'biggest kind' of fowls!

"The group represents this variety with accuracy, and are, without doubt, for their kind, rare specimens of the genuine *gallus giganteus* of modern ornithologists. As Her Majesty has long been known among the foremost patrons of that agreeable branch of rural pursuits, poultry-raising, we do not doubt but that this splendid present from Mr. Burnham will prove highly gratifying to her tastes in this particular."

Portraits of these fowls appeared in *Gleason's Pictorial* for January, 1853, and the editor spoke as follows of them:

"The Grey Shanghae Fowls lately presented to Her Majesty Queen Victoria, of Great Britain, by George P.

Burnham, Esq., of Boston, were extraordinary specimens of domestic poultry, and were bred the past season by Mr. Burnham from stock imported by him direct from China. They were universally admitted, by the thousands who saw them before they left, to be the largest and choicest-bred lot of chickens ever seen together in this vicinity. These fowls were from the same broods as those lately sent to Northby, of Aldborough, by Mr. Burnham, who is, perhaps, the most successful poultry-raiser in America; and while these beautiful birds are creditable to him as a breeder, they are a present really 'fit for a queen.'"

The New York journals alluded to them in flattering terms, during their transit through that city on the way to their destination; and the numerous orders that crowded in upon me was the best evidence of the estimation in which this variety of domestic fowls was then held, as well as of the determined disposition of "the people" to be supplied from my "*pure*-bred stock."

By one of the British steamers, in the summer of 1853, the express of Edwards, Sanford & Co., took out to Europe from my stock, for Messrs. Bakers, of Chelsea, Baily, of London, Floyd, of Huddersfield, Deming, of Brighton, Simons, of Birmingham, and Miss Watts, Hampstead, six cages of these "extraordinary" birds. The best of the hens weighed nine to nine and a half pounds each, and three of the cocks drew over twelve pounds each! There were forty-two birds in all, which, together, could not be equalled, probably, at that time, in America or England, for size, beauty and uniformity of color. The sum paid me for this lot of Greys was eight hundred and seventy dollars.

Of the three fowls sent to Mr. John Baily (above mentioned), and which he exhibited in the fall of that year in England, the following account reached me, subsequently:

"Mr. Geo. P. Burnham, of Melrose, sent out to England, last fall, to Mr. John Baily, of London, a cage of his fine 'Grey Shanghaes,' which were exhibited at the late Birmingham Show. The London *Field* of Dec. 24th says that '*one pair* of these fowls, from Mr. Burnham, of the United States, the property of Mr. Baily, of Mount-street, were shown among the extra stock, and were purchased from him, during the exhibition, by Mr. Taylor, of Shepherd's Bush, at one hundred guineas' ($500)!"

This was the biggest figure ever paid for *two* fowls, I imagine! Mr. Baily paid me twenty pounds sterling for the trio, and I thought that fair pay, I remember. The following brief account of my trade for the year of our Lord 1853, I published on the last day of December of that year, for the gratification of my numerous friends, and for the information of "the people" who felt an interest in this still exciting and (to me) very agreeable subject:

"EDS. BOSTON DAILY TIMES: In a late number of your journal you were pleased to allude to the sales of live-stock made by me latterly. At the close of the present year, I find upon my books the following aggregate of sales for 1853, which — to show how much has been done by *one* dealer — may be interesting to some of your readers who 'love pigs and chickens.'

"I have sent into the Southern and Western States, through Adams & Co.'s Express alone, from Jan. 1st to Dec. 27th, 1853, a little rising $17,000 worth of Chinese fowls and fancy pigs. By Edwards, Sanford & Co.'s Transatlantic Express, in the same period, I have sent to England and the continent about $2000 worth of my 'Grey Shanghaes.' By Thompson and Co. and the American Western Express Co., I have sent west and south-west, in the same time, over $1200 worth; and my minor cash sales (directly at my yards in Melrose) have been over $1000; making the entire sales from my establishment for the past year nearly or quite *twenty-two thousand dollars* in value. Of this amount, $7300 worth has been sold since the 10th of Sept. last.

"By the first steamer that leaves New York in January, '54, I shall send to New Orleans (to a single customer) between five and six hundred dollars' worth, ordered a few days since. I have also now in hand three large orders to fill for Liverpool and London, immediately; and the present prospect is that the poultry-trade will be considerably better next year than we have ever yet known it in New England. Wishing you and my competitors in the trade a "Happy New Year,' I am theirs and yours, truly,

"GEO. P. BURNHAM.

" *Melrose, Dec.* 30, 1853."

I have offered these statistics and facts to give some idea of the amount of trade that must have been current, in the *aggregate*, when these isolated instances are considered, and for the purpose of affording the reader an opportunity to judge measurably to what an extent this *fever* really raged.

Thousands and tens of thousands of "the people" were now (or had been) engaged in this extraordinary excitement, who were continuously humbugging themselves and each other, at round cost. And when these thousands are multiplied by the fives or tens, twenties or fifties, one hundreds or five hundreds of dollars, that they invested in this mania, the "prime cost" of this hum can be fancied, though it can never be known with accuracy.

CHAPTER XXI.

EXPERIMENTS OF AMATEURS.

THE newspapers of the day were now occupied with speculative and actual statistics, of various kinds, relating to the utility and value of poultry and its produce, and every one seemed to join, in his or her way, to magnify the vastness of this enterprise; and statements like the following, in respectable public journals, had the effect to increase and keep up to fever-heat the state of the hen malady:

"By reference to the agricultural statistics of the United States, published from reliable sources in 1850, it may be seen that the actual value of poultry, in New York State alone, was two millions three hundred and seventy-three thousand and twenty-nine dollars! Which was more than the value of *all the swine* in the same state; nearly equal to *one half the value of its sheep*, the *entire* value of its *neat cattle*, and nearly *five times* the value of its *horses and mules!*"

The amount of sales of live and dead *poultry* in Quincy Market, Boston, for the year 1848, said another paper, was six hundred seventy-four thousand four hundred and twenty-three dollars: the average sales of one dealer alone amounting to twelve hundred dollars per week for the whole year. The amount of sales for the whole city of Boston, for the same year, was over one million of dollars. The amount of sales of *eggs* in and around the Quincy Market for 1848 was one million one hundred and twenty-nine thousand seven hundred and thirty-five dozen, which, at eighteen cents per dozen, makes the amount paid for eggs to be two hundred three thousand three hundred and fifty-two dollars and thirty cents; while the amount of sales of eggs for the whole city of Boston, for the same year, was a fraction short of one million of dollars; the daily consumption of eggs at one of its hotels being seventy-five dozen daily, and on Saturday one hundred and fifty dozen.

At this time, a single dealer in the egg-trade, at Philadelphia, sent to the New York market, daily, one hundred barrels of eggs; while the value of eggs shipped from Dublin to Liverpool and London was more than five millions of dollars for the year 1848.

In addition to these facts, frequent allusions were made to the enormous quantities required for other markets, in the interior, to supply which the number of laying hens must be kept good, and increased, as the demand for the eggs was constantly augmenting, and the business, "if skilfully and judiciously managed" (said the agricultural papers), *must* prove immensely profitable to those who engage in it.

If "skilfully and judiciously managed"! This was good advice. But no one could inform "the people" how this management was to be effected. In the mean time, every sort of experiment was resorted to, by amateurs and fanciers and humbugs (who had been humbugged), to "improve" the breeds of poultry, and to produce new fowls that would lay two or three or four eggs for one, as compared with the old-fashioned birds.

We knew one beginner who had purchased a pretty little place a few miles from the city, who contracted the fever, and "suffered" badly, but who was cured by the following curious result of his early experiments. Eggs were scarce (genuine ones), and, after considerable searching, he finally procured of some one in Boston a clutch of "fancy" eggs, for which he paid big figures, but which did not *turn out* exactly what he anticipated; and so *he* concluded, after a time, that the hen fever was a rascally hum. (He did n't procure these eggs of *me*, be it understood. *I* never had any but *genuine* ones!)

He purchased what he was assured were pure "Cochin-China" eggs. (Perhaps they were — who knows?) And after waiting patiently for six long weeks for the "curious" eggs to hatch, he found six young *ducks* in his coop, one morning! — So much for his knowledge of eggs!

But this was not so bad as was the case of one of his neighbors, however, who paid a round price for half a dozen choice eggs, queer-looking speckled eggs — small, round, "outlandish" eggs — which he felt certain would produce *rare* chicks, and which he was very cautious in setting under his very best hen.

At the end of a few days he was startled, at the breakfast-table, to hear his favorite hen screaming "bloody murder" from within the coop! He rushed to the rescue, raised the box-lid, and found her still on the nest, but in a frightful perturbation — struggling, yelling and cackling, most vociferously.

He spoke to her kindly and softly; he would fain appease and quiet her; for there was great danger lest, in her excitement and struggles, she would destroy the favorite eggs — those rare eggs, which had cost him so much money and trouble. But soft words were vain. His "best" hen continued to scream lustily, and he raised her from the nest to look into the cause of the trouble more critically. His astonishment was instantaneous, but immense; and his surprise found vent in the brief but expressive exclamation, "*Turkles — by thunder!*"

Such was the fact. This poor, innocent poultry-"fancier" was the victim of misplaced confidence. The party who sold him *them* eggs had sold the buyer shockingly! And instead of a brood of pure Cochin-Chinas, he found that his favorite hen had hatched half a dozen pure *mud-turtles*, all of which, upon breaking from the shells, seized upon the flesh of the poor fowl, and had well-nigh taken her life before they could be "choked off." He has given up the chicken-trade, and has since gone into the dwarf-pear business. Poor devil!

A youthful lawyer of my acquaintance, away Down East, who was proverbial for his "sharp practice" at the bar, met with a young doctor, who was a great bird-fancier, and with whom he subsequently formed an intimate acquaintance. Our medicinal friend owned a pretty little estate, distant a few miles from the city of P——, where he kept up a very neat establishment, which was thoroughly appointed. Among his out-of-door appurtenances, he maintained a modern bee-house, a choice dove-cot, and a well-selected aviary; in the latter he had some choice poultry, and into this the doctor invited his legal associate, one day, to examine his specimens of cacklers and crowers.

There was a super-excellent "Bother'em" fowl among this collection,—a rare hen, the many good qualities of which the doctor dilated on (as he always did before his visitors), and the lawyer took a fancy to the beauty, instanter; but this fowl was a great favorite, and the doctor would neither sell, lend, or give her away; and then the visitor begged some of her eggs, as a last favor. But the doctor was selfish in regard to this particular bird — he wanted the breed exclusively to himself. It was of no avail, however, and his friend promised to embrace the first opportunity to steal the hen, and all the eggs he could find, if his request were not complied with; whereupon the doctor at length reluctantly promised to send him a dozen within a week, provided he said nothing about it. He would do it for *him*, as a particular favor — and so he was as good as his word.

The young lawyer had his poultry-yard, also; and, selecting a fine hen, he quickly set her upon the choice Bother'em eggs, resolved to have as good a show as his neighbor. But three weeks passed — four, and upwards — but no chickens appeared! He broke up the nest, at last, and then called upon the doctor at once.

"What luck, Tom?"

"Not a chick!"

"No!"

"Not a *one*. The eggs were n't good."

"No? That is queer," continued the doctor, "when I took so much extra pains with 'em."

"Extra pains — how?"

"Why, *I boiled every one of 'em*, the last thing before I sent 'em down to you!"

And so he did. Tom grinned, squirmed, and went home, — but that was n't the last of this joke.

Six months afterwards, the keen-witted doctor visited the

lawyer's little place, where he saw a magnificent large Bucks County rooster stalking about in the latter's yard.

"By Jove, Tom! That 's a rouser," exclaimed the doctor, enthusiastically, "'pon my word! Where d' you get him?"

"Pennsylvania — Buxton's; a fine fellow that. Only eight months old."

"Will you sell him?"

"Yes — no; I reckon not, on the whole."

"I 'll give you an X for him."

"Well, take him. He 's worth twenty dollars; but you shall have him for ten dollars, being an old friend."

The doctor placed the huge crower in his gig immediately, went home, killed off two of the finest Dorking roosters in the county, and put the new comer into his nice poultry-house; congratulating himself upon having at last secured a "tip-top breeder," and nothing else.

At the end of the season, however, he complained to his friend the lawyer that he had had but very few eggs latterly; he could raise no chickens from them — not a *one;* and he did n't think much of the ten-dollar bird he purchased of him, any way.

"He 's a rouser, Bill, surely," said the lawyer, with a knowing smirk, repeating the doctor's exclamation on first beholding the rooster.

"Well, yes — large, large — but — "

"And a finer *capon* I never sold to anybody in my life!"

"A *what!*" screamed the doctor, springing towards his horse, which stood near by.

"What's the price of *b'iled eggs*, Bill?" roared the lawyer, in reply.

"Ten dollars a dozen, by thunder!" was the answer, as the doctor drove his rowels into the sides of his nag, and dashed away from his friend's gate a *wiser* if not a better man.

Many amateur poultry-raisers resorted to the most ridiculous and injurious shifts for remedies against the ills that hen-flesh is heir to. I have known certain friends who passed two or three hours every morning in running about their fowl-premises with pill-box and pepper-cup in hand, zealously dosing their drooping chickens, to their certain destruction. And some of the "doctors" went into *jalap*, in cases of colds, fevers, &c., in their fowls. We should as soon think of using arsenic, or any other poison, under such circumstances. The internal formation of a hen is scarcely believed to resemble that of a human being, surely; and why such medicinal applications, pray? This reminds us of a private joke, by the way, that was "let out" by a young fancier (out West) a little while ago.

He had a bad cold himself, and had mixed "summat hot" to swallow, one evening. His servant informed him that his favorite Cochin-China crower had been ill for a day or two; and he ordered twenty grains of jalap to be prepared for his fine bird. By some mistake his toddy was given to the crower, and he swallowed the hen-medicine himself, and retired to bed.

He slept soundly for a time, but was visited with shocking dreams. He fancied himself to be a huge rooster—one of the biggest kind; that he had taken all the premiums at all the shows, and that he had finally been set to hatch over a bushel of Shanghae eggs. It was the twentieth day, at last, and the chickens commenced to come forth from their shells beneath him. He dare not move,—his fowl-cure was at work,—and his critical position, for the time being, can be better imagined than portrayed. With a desperate effort, and a shrieking crow, he at length sprang from his couch, dashed out of doors, and, since the day afterwards, has resolved to eschew the use of jalap among his poultry,—a determination which, in all candor, we recommend earnestly to the hen-Galens who imagine that a hen is "a human."

It had now become an every-day occurrence to hear of black chickens emerging from what were "warranted" pure white fowls' eggs; top-knot birds peeped forth from the eggs of pure-bred anti-crested hens; and all colors and shapes and varieties of chickens, except those that they were purchased for, made their appearance about the time of hatching the eggs so bought.

All the old-fashioned fowls were utterly discarded. Cochin-Chinaism, Shanghaeism, Bother'em Pootrumism, was rampant. The fancy egg-trade had begun to fall off sensibly. "The people" had had enough of *this* part of the enterprise, which was destined to prove so "immensely profitable," if "judiciously and skilfully managed;" and the price was reduced to the miserable sum of three to five dollars a dozen, only, as customers chanced to turn up.

From the commencement of the trade, in 1849, down to the month of August, 1853, I had a continued and certain sale, however, for every egg deposited upon my premises, at *my* price.

But this, though an exception, was not to be wondered at. *I* kept and raised only the "genuine" article.

CHAPTER XXII.

TRUE HISTORY OF "FANNY FERN."

I WAS riding through Brookline, Mass., one fine afternoon, on my round-about way home from a fowl-hunting excursion in Norfolk County, when my attention was suddenly attracted by the appearance and carriage of the most extraordinary-looking bird I ever met with in the whole course of my poultry experience.

I drew up my horse, and watched this curiosity for a few minutes, with a fowl-admirer's wonder. It was evidently a *hen*, though the variety was new to me, and its deportment was very remarkable. Her plumage was a shiny coal-black, and she loitered upon a bright-green bank in the sunshine, at the southerly side of a pretty house that stood a few yards back from the road. She was rather long-legged, and "spindle-shanked," but she moved about skippingly and briskly, as if she were treading upon thin egg-shells. Her feet were very delicate and very narrow, and her body was thin and trim; but her plumage — that glossy, jet-black, brilliant feathery habit — was "too much" for my then excited "fancy" for beautiful birds; and I thought I had never seen a tip-top fowl before.

As I gazed and wondered, this bird observed me coquettishly, and, raising herself slightly a tip-toe, she flapped her bright wings ludicrously, opened her pretty mouth, and sent forth a *crow* so clear and sharp, and so utterly defiant and plucky, that I laughed outright in her face. I did. I could n't help it.

She noticed my merriment, and instantly flap went those glittering wings again, and another shout — a very shriek of a crow, a termagant yell of a crow — rang forth piercingly from the lungs of my sable but beautiful inamorata.

This second crow was full of fire, and daring, and challenge, and percussion. It seemed to say, as plainly as words could have uttered it, "Who are *you?* What you after? Would n't you like to cage *me* up — *s-a-y?*"

I laughed again, wondered more, stared, and shouted "Bravo! Milady, you *are* a rum 'un, to be sure!" And again she hopped up and crowed bravely, sharply, maliciously, wildly, marvellously.

I was puzzled. I had heard of such animals before. I had read in the newspapers about Woman's Rights conventions. I had seen it stated that hens occasionally were found that "crowed like a cock." But I had never seen one before. This *was* an extraordinary bird, evidently.

There it went again! That same shrill, crashing, challenging crow, from the gullet of the ebon beauty before me. O, *what* a crow was that, my countrymen! I resolved to possess this bird, at any cost. And I was soon in communication with the gentleman who then had her.

"Is this *your* hen, sir?" I inquired. And I think the gentleman suspected me, instanter.

"Yes," he answered. "That is, I support her."

"Will you sell her?"

"No — no, sir."

"I will give you ten dollars for her."

Crack! Crash! Whew! went that crow, again. I was electrified.

"I'll give you fifteen ——"

"No, sir."

"Twenty dollars, then."

"No."

"What will you take for her?"

"Hark!" he replied. "Is n't that music? Is n't that heavenly?"

"What *is* that?" I asked, eagerly.

"My hen."

"What is she doing?"

"Singing," said the gentleman.

"Beautiful!" I responded. "I will give you forty dollars for her."

"Take her," replied her keeper. "She is yours."

"What breed is it?" I inquired.

"Spanker," said the gentleman, "but rare. It is one of Ellett's importation — genuine."

"Remarkable pullet!" I ventured.

"Hen, sir, *hen*," insisted the stranger.

I paid him forty dollars down, and seized my prize, though she proved hard to catch.

"She 's much like the Frenchman's flea, sir," said her previous possessor. "Put your finger on her, and she 's never there. Feed her well, however, keep her in good quarters, let her do as she pleases, and she 'll always crow — always, sir. Hear *that?* You can't stop her, unless you stop her breath. She always crows and sings. There it is again! Is n't that a crow, for a hen — eh?"

It was, indeed.

"Good-day," said the Brookline gentleman, quietly pocketing his money. "Fanny will please you, I 've no doubt."

"Fanny?" I queried.

"Yes; I call her '*Fanny Fern*,'" said the stranger to me, as I entered my wagon; and, half an hour afterwards, my forty-dollar cock-hen, "Fanny Fern," was crowing again furiously, lustily, magnificently, on the bright-green lawn beneath my own parlor-windows.

"Fanny" proved a thorough trump. Bantams, Games, Cochins, Dorkings, Shanghaes, Bother'ems, were *no*where when "Fanny" was round. She could outcrow the lustiest feathered orchestra ever collected together in Christendom. She was a wonder, that redoubtable but frisky, flashy, sprightly, sputtery, spunky "Fanny Fern."

And did n't the boys run after her? Well, they did! And did n't they want to buy her? Did n't they bid high for her, at last? Did n't everybody flock to see her, and to hear "Fanny" crow? And *did n't* she continue to crow, too? Ah! it was heaven, indeed (and sometimes the other thing), to listen to "Fanny's" voice.

When "Fanny" opened her mouth, everybody held their breath and listened. "Fanny" crowed to some purpose, verily! She crowed lustily against oppression, and vice, and wrong, and injustice; and she crowed aloud (with her best strength) in behalf of injured innocence, and virtue, and merit, exalted or humble.

And, finally, "Fanny" hatched a brace of chickens; and *did n't* she crow for and over *them?* She now cackled and scratched, and crowed harder and louder and shriller than ever. The people stopped in the street to listen to her; old men heard her; young men sought after her; all the women began to "swear" by her; the children thronged to see her; the newspapers all talked about her; and thousands of books were printed about my charming, astonishing, remarkable, crowing "Fanny Fern."

I sent her to the fowl-shows, where she "took 'em all

down" clean, and invariably carried away the first premium in her class. Never was such a hen seen, before or since. I was offered a hundred, two hundred, five hundred dollars for her. I was poor; but did n't I own this hen "Fanny,"—the extraordinary, wonderful, magnificent, coal-black, blustering, but inapproachable and world-defying "Fanny"?

"I will give you *eight* hundred dollars for her," said a publisher to me, one day. "I want to put her in a book. She 's a wonder! a star of the first magnitude! a diamond without blemish! a God-send to the world in 1854!"

At this moment "Fanny" crowed.

"Will you take eight hundred?" screamed the publisher, jumping nearly to the ceiling.

"No, sir."

"A thousand?"

"No."

"Two thousand?"

"No, sir."

"*Five* thousand?"

"No! I will keep her."

And I did. What was five thousand dollars to me? *Bah!* I had the hen-cock "Fanny Fern." I did n't want money. My pocket-book was full to bursting, and so was my head with the excitement of the hen fever. And "Fanny" crowed again. Ah! *what* a crow was Fanny's!"

"Fanny" could n't be bought, and so my competitors clanned together to destroy her. The old fogies did n't like this breed, and they resolved to annihilate all chance of its perpetuation. I placed her in better quarters, where she would be more secure from intrusion or surprise. I told her of my fears,—and *did n't* she crow? She flapped her bright black wings, and crowed all over. "Cock-a-doodle-*doo — oo — oo!*" shouted "Fanny," while her sharp eyes twinkled, her fair throat trembled, and the exhilarating tone of defiance seemed to reach to the very tips of her shining toe-nails. "Cock-a-too —*roo — oo!*" she shrieked; "let 'em come, too! See what they 'll *do—oo!* I 'll take care of *you—oo!* Don't get in a *stoo—oo!* Pooh—pooh — poo — *poo!*"

Maybe "Fanny" did n't crow! And *I* learned to crow. It was beautiful! She crowed, and I crowed. We crowed together. She in her way,—I in mine. The duet was mellifluous, cheering, soul-stirring, life-invigorating, *profitable.*

"Fanny" went into New York State, crowing when she left, crowing as she went, and continuing to crow until she crowed the community there clear through the next fourth o' July, out into the fabled millenium. She crowed Messrs. Derby & Miller into a handsome fortune, and Mason & Brothers into ditto. She crowed one Hyacinth into the shreds of a cocked hat and battered knee-buckles. She crowed the Hall breed of old hens so far out of sight that the "search for Sir John Franklin" would be a fool to the journey requisite to overtake that family. And still she *crowed.*

The more they bade her stop, the more she would n't. "Cock-a-tootle — *too!*" "I-know-what-*I*-shall — *doo!*" "What-do-I-care-for — *yoo?*" "This-world-is-all — foo-*foo.*" "Leave-*me*-and-I'll-leave — *you.*" "If-not-I'll-lamm — *you* — TOO — OO!"

And "Fanny" crowed herself at last into the good graces of two *long brothers* in Gotham, where she is now crowing with all her might and main. Let her crow!

She was a remarkable "bird," that rollicking, joyous, inexplicable, flirting, funny, furious "Fanny Fern." I hear her now again!

"COCK-A-DOODLE — DOO — OÖ!" "Young 'Un,— you-will-do!!" "*Et — tu — Brute — o-o-o!!!*"

CHAPTER XXIII.

CONVALESCENCE.

ONE striking feature that exhibited itself in the midst of this mania, was the fact that prominent among the leading dealers in fancy poultry, constantly appeared the names of clergymen, doctors, and other "liberally-educated" gentlemen.

In Ohio, Pennsylvania, New York, and most of the Eastern States, this circumstance was especially noticeable; and more particularly in England. Whether this class of the community had the most money to throw away, or whether their leisure afforded them the better opportunity to indulge in this fancy, I cannot say; but one thing is certain,—among my own patrons and correspondents, for the past five or six years, I find the names of this class of "the people" by far the most conspicuous and frequent.

There came into my office, one morning late in 1853, a Boston physician (whom I had never seen before), who introduced himself civilly, and invited me to ride a short distance with him up town. I was busy; but he insisted, and his manner was peculiarly urgent and determined.

"My carriage is at the door," he said; "and I will bring you back here in twenty minutes. I have some pure-blood stock I desire to dispose of."

"What *is* it, doctor?" I asked.

"Chickens, chickens!" replied the doctor, briefly.

I assured the gentleman that I had near a thousand fowls on hand at this time, and had no possible wish to increase the number.

"They are pure-bred — cost me high," he continued; "are very fine, but I must part with them — come!"

I joined him, and we rode a mile or more, when he halted before a fine, large house; his servant in waiting took his horse, and he ushered me into his well-appointed poultry-house, at the rear of his dwelling.

The buildings were glazed in front and upon the roofs; the yards were spacious and cleanly, and appropriately divided; the laying and hatching rooms were roomy and convenient; the roosting-house was airy and pleasant, and everything was, seemingly, in excellent order, and arranged with good taste throughout.

"That cock cost me twenty dollars," said the doctor, calmly. "Those two hens I paid eighteen dollars for. That bird, yonder, twelve dollars. These five pullets stand me in about forty-five dollars. I have never yet been able to hatch but one brood of chickens. The rats carried *them* off by the third morning after they came into this world. The hens sometimes lay, I believe; at least, my man says so. I have never *seen* any eggs from them myself, however. I have no doubt this species of fowls (these Changays) *do* lay eggs, though. There are twenty-two of them. Buy them, Mr. B——," continued the doctor, urgently.

I said no; I really did not want them.

"I *had* nigh forty of them," continued the doctor, "two months ago. But they have disappeared. Disease, roup, vermin, night-thieves, sir. Will you buy them? John —— drive them out!"

The fowls were driven into the main yard. There were but sixteen in all.

"Where are the rest, John?" inquired the doctor, anxiously. "There were twenty-two here yesterday."

"I dunno, sir," said John.

"Drive 'em back, and box them up, John. Mr. B——, will you make an offer for the remainder? To-morrow I shall probably have none to sell! Will you give anything for them?"

I declined to buy.

"Will you permit me to send them to you as a present, sir?" he continued.

I did not want them, any way. I had a full supply.

"What will you charge me, Mr. B——, to allow them

to be sent to you?" continued the fancier, desperately, and resolutely, at last.

I saw he was determined, and I took his fowls (fifteen of them), and gave him ten dollars.

He smiled.

"I have had the hen fever," he added, "*badly* — but I am better of it. I am convalescent, now," said the doctor. "You see what I have here for houses; cost me over seven hundred dollars; my birds over four hundred more; grain and care for a year, a hundred more. I am *satisfied!* Your money, here, is the first dollar I ever received in return for my investment. You see what I have left out of my venture of twelve or thirteen hundred dollars; the manure, and — and — the lice!"

Such were the exact facts! His stock was selected from the Marsh and Forbes importations, and the birds were good; but, by the time he got ready to believe that it was n't *all* gold that glittered, the sale of *this* variety of fowl had passed by. A chance purchaser happened to come along soon after, however, who "had n't read the papers" so attentively as some of us had, and who wanted these very fowls. I sold them to him, "cheap as a broom," because the fever for this kind of bird was rapidly declining. He paid me only $150 for this lot; which *was* a bargain, of a truth. The buyer was satisfied, however, and so was *I*.

These were but isolated instances. Scores and hundreds

of gentlemen and amateur fanciers found themselves in a similar predicament, at the end of one or two or three years. Without possessing a single particle of knowledge requisite to the successful accomplishment of their purpose,— utterly ignorant of the first rudiments of the business,— they jumped into it, without reason, forgetting the wholesome advice contained in the musty adage, "look before you leap." And, after sinking tens and hundreds or (in some cases) *thousands* of dollars in experiments, they woke up to find that they had *had* the fever badly, but, fortunately, were at last convalescent!

I was busy, all this time, in supplying my friends with "pure-bred" stock, however, and had very little leisure to tarry to sympathize with these "poor crecturs." The demand for *my* stock continued, and the best year's business I ever enjoyed, was from the spring of 1853 to May and June, 1854; when it commenced to fall off very sensibly, and the prospect became dubious, for future operations, even with *me*.

CHAPTER XXIV.

AN EXPENSIVE BUSINESS.

DURING the past six years I have expended, outright, for breeding stock, and for appropriate buildings for my fowls, over four thousand dollars, in round numbers — without taking into the account the expenses of their care, and the cost of feeding.

Few breeders have spent anything like this sum, for this purpose, *strictly*. In the mean time, the aggregate of my receipts has reached (up to January, 1855) upwards of seventy thousand dollars. I have raised thousands upon thousands of the Chinese varieties of fowls, and my purchases to fill orders which came to hand during this term — in addition to what I was able to fill from those I myself raised — have been very large. And, while I have been thus engaged, hundreds and hundreds of amateurs and fanciers have sprung up in various directions, all of whom have had their share, too, in this trade.

To the fanciers — those who purchased, as many did at first, simply for their amusement, or for the mere satisfaction of having good, or, perhaps, the best birds—this fever proved an expensive matter. I have known amateurs who willingly paid twenty, fifty, or a hundred dollars, and even more, for a pair, or a trio, of what were considered very choice Shanghaes. These fowls, after the first few weeks or months of the purchaser's excitement had passed by, could be bought of him for five or ten dollars a pair! Yet, his next-door neighbor, who would not now take these identical birds for a gift, scarcely, would pay to a stranger a similarly extravagant amount to that which had a hundred times been paid by others before him, for something, perhaps, inferior in quality, but which chanced to be called by the most popular name current at the moment.

Thus, for a time, bubble number one, the *Cochin-Chinas*, prevailed. The eggs of these fowls sold at a dollar each, for a long period. Then came the *Shanghaes*, of different colors,—as the yellow, the white, the buff, or the black,—and took their turn. Many thousands of these were disposed of, at round rates. The smooth-legged birds at first commanded the best price; then the feathered-legged. And, finally, came the Grey Shanghaes, or "Chittagongs," or "Brahmas," as they were differently termed; and this proved bubble number two, in earnest.

Everybody wanted them, and everybody had to pay for them, too! They were large, heavy fowls, of China blood, plainly, but, with some few exceptions, were indifferent birds. They were *leggy*, however, and stood up showy and tall, and, to *look at*, appeared advantageously to the fancy, at this period. In the maw of this bubble, thousands of good dollars were thrown; and no race of poultry ever had the run that did these Greys, under various names, both in this country and in England.

A most excellent Southern trade had sprung up, and large shipments of fowls went forward to the West, from Massachusetts, and to Charleston, Augusta, Mobile, New Orleans, etc., where the fever broke out furiously, and continued, without abatement, for three years or more.

No buyers were so liberal, generally, and no men in the world, known to Northern breeders, bought so extensively, as did these fanciers in New Orleans and vicinity. They purchased largely, from the very start; and the trade was kept up with a singular vigor and enterprise, from the beginning to the end. Orders, varying in value from $500 to $1200 and $1500, were of almost weekly occurrence from that region; and in one instance, I sent forward to a gentleman in Louisiana, a single shipment for which he paid me $2230! This occurred in September, 1853.

In this same year, I sent, from January to December, to another gentleman (at New Orleans), over *ten thousand* dollars' worth of stock.

The prices for chickens ranged from $12 @ $15 a pair, to $25 or $30, and often $40 to $50, a pair. These rates were always willingly and freely paid, and the stock was, after a while, disseminated throughout the entire valley of the Mississippi; where the China fowls always did better than in our own climate.

It proved an expensive business to some of these gentlemen, most emphatically. But they always paid cheerfully, promptly, and liberally; and *knew* the Yankees they were dealing with, a good deal better than many of the sharpers supposed they did. For myself, I shall not permit this opportunity to pass without expressing my thanks to my numerous and generous Southern patrons, to whom I sent a great many hundred pairs of what were deemed "good birds," and to whom I am indebted, largely, for the trade I enjoyed for upwards of five years. I sincerely hope they made more money out of all this than I did; and I trust that their substance, as well as "their shadows, may never be less."

During this year, and far into 1854, the current of trade turned towards Great Britain; and John Bull was not very slow to appreciate the rare qualities of my "magnificent" and "extraordinary" birds; "the like of which," said a London journal, when the Queen's fowls first arrived, "was never before seen in England."

For upwards of a year, I had all *this* trade in my own way. Subsequently, some of the smaller dealers sent out a few pairs to London, but "the people" there could never be brought to believe those fowls were anything but mongrels; and, while these interlopers contrived to murder the trade there, they at the same time "cut off their own noses," for the future, with those who knew what poultry was, upon the other side of the Atlantic.

I had *my* shy at the Britons, seasonably!

But, a few months afterwards (as I shall show in a future chapter), through the mismanagement of an ambitious dealer in other fancy live-stock, the trade with England, from this side of the water, was completely ruined. Over two hundred American fowls were thrown suddenly upon the London market, and were finally sold there, at auction, for a very small sum; and we were subsequently unable (with all our chicken-eloquence) to make John Bull believe that even the *Grey Shanghaes* were any longer "scarce" with us, here!

CHAPTER XXV.

THE GREAT PAGODA HEN.

THE most ridiculous and fulsome advertisements now occupied the columns of certain so-called agricultural papers in this country, particularly one or two of these sheets in New York State.

Stories were related by correspondents (and endorsed by the nominal editors), regarding the proportions and weights and beauties of certain of the "Bother'em" class of fowls, that rivalled Munchausen, out and out. Fourteen and fifteen pound cocks, and ten or eleven pound hens, were as common as the liars who told the stories of these impossibilities. And one day the following capital hit, by Durivage, appeared in a Boston journal. He called it "The Great Pagoda Hen." There is as much truth in this as there was in many of the more seriously-intended articles of that time. It ran as follows:

"Mr. Sap Green retired from business, and took possession of his country 'villa,' just about the time the 'hen fever' was at its height; and he soon gave evidence of having that malignant disorder in its most aggravated form. He tolerated no birds in his yard that weighed less than ten pounds at six months, and he allowed no eggs upon his table that were not of a dark mahogany color, and of the flavor of pine shavings. He supplied his own table with poultry, and the said poultry consisted of elongated drumsticks, attached by gutta-percha muscles and catgut sinews to ponderous breast-bones. He frequently purchased a 'crower' for a figure that could have bought a good Morgan horse; but then, as the said crower consumed as much grain as a Morgan horse, he could not help being perfectly satisfied with the bargain. His wife complained that he was 'making ducks and drakes' of his property; but, as that involved a high compliment to his ornithological tastes, he attempted no retort. He satisfied himself that it 'would pay in the end.' His calculations of profits were 'clear as mud.' He would have a thousand hens. The improved breeds were warranted to lay five eggs apiece a week; and eggs were worth — that is, *he was paying* — six dollars a dozen. His thousand hens would lay twenty thousand eight hundred and thirty-three dozen eggs per annum, which, at six dollars per dozen, would amount to the sum of one hundred and twenty-four thousand nine hundred and ninety-eight dollars. Even deducting therefrom the original cost of the hens and their keep,— say thirty-six thousand dollars,— the very pretty trifle of eighty-eight thousand nine hundred and ninety-eight was the remainder — clear profit. Eggs — even dark mahogany eggs — *went down to a shilling a dozen!* But we will not anticipate.

"To facilitate the multiplication of the feathered species, Mr. Green imported a French Eccaleobion, or egg-hatching machine, that worked by steam, and was warranted to throw off a thousand chicks a month.

"One day an 'ancient mariner' arrived at the villa, with a small basket on his arm, and inquired for the master of the house. Sap was just then engaged in important business,— teaching a young chicken to crow,— but he left his occupation, and received the stranger.

"'Want to buy an egg?' asked the mariner.

"'One egg? Why, where did it come from?' asked the hen-fancier.

"'E Stingies,' replied the mariner.

"'Domestic fowl's egg?'

"'Domestic.'

"'Let 's see it.'

"The sailor produced an enormous egg, weighing about a pound. Sap 'hefted' it carefully.

"'Did you ever see the birds that lay such eggs?' he asked.

"'Lots on 'em,' replied the sailor. 'They 're big as all out-doors. They calls 'em the Gigantic Pagoda Hen. I 'm afeared to tell you how big they are; you won't believe me. But jest you hatch out that 'ere, and you 'll see wot 'll come of it.'

"'But they must eat a great deal?' said Sap.

"'Scarcely anything,' replied the mariner; 'that 's the beauty on 'em. Don't eat as much as Bantams.'

"'Are they good layers?'

"'You can't help 'em laying,' replied the seaman, enthusiastically. 'They lay one egg every week-day, and two Sundays.'

"'But when do they set?' queried Green.

"'They don't set at all. They lays their eggs in damp, hot places, and natur' does the rest. The chicks take keer of themselves as soon as they 're out of the shell.'

"'Damp, hot place!' said Sap. 'My Eccaleobion is the very thing, and my artificial sheep-skin mother will bring 'em up to a charm. My friend, what will you take for your egg?'

"'Cap'n,' said the mariner, solemnly, 'if I was going to stay ashore, I would n't take a hundred dollars for it; but, as I 've shipped ag'in, and sail directly, you shall have it for forty.'

"The forty dollars were instantly paid, and the hen-fancier retired with his prize, his conscience smiting him for having robbed a poor, hard-working sailor.

"O, how he watched the egg-hatching machine while that extraordinary egg was undergoing the steaming process! He begrudged the time exacted by eating and sleeping; but his vigils were rewarded by the appearance, in due time, of a stout young chick, with the long legs that are a proof of Eastern blood. The bird grew apace; indeed, almost as rapidly as Jack's bean-stalk, or the prophet's gourd. But the sailor was mistaken in one thing; it ate voraciously. Moreover, as it increased in size and strength, the Pagoda exhibited extraordinary pugnacity. It kicked a dozen comrades to death in one night. It even bit the hand of the feeder. Soon it was necessary to confine it in a separate apartment. Its head soon touched the ceiling. What a pity it had no mate! Sap wrote to a correspondent at Calcutta to ship him two pairs of the Great Pagoda birds, without regard to cost. Meanwhile he watched the enormous growth of his single specimen. He kept its existence a profound secret. It was under lock and key, in a separate apartment, lighted by a large window in the roof. Sap's man-of-all-work wheeled daily two bushels of corn and a barrel of water to the door of the apartment, and Green fed them out when no one was looking. Even this supply was scanty; but, out of justice to his family, Sap was compelled to put the monster bird on allowance.

"'Poor thing!' he would say, when he saw the creature devouring broken glass, and even bolting stray nails and gravel-stones, 'it cuts me to the soul to see it reduced to such extremity. But it's eating me out of house and home.

15

Decidedly, that sailor-man must have been deceived about their being moderate feeders.'

"When the bird had attained to the enormous altitude of six feet, the proud proprietor sent for the celebrated Dr. Ludwig Hydrarchos, of Cambridge, to inspect him, and furnish him with a scientific description, wherewith he might astonish his brethren of the Poultry Association. The doctor came, and was carefully admitted by Green to the presence of the Great Pagoda Hen. The bird was not accustomed to the sight of strangers, and began to manifest uneasiness and displeasure at seeing the man of science. It lifted first one foot and then the other, as if it were treading on hot plates.

"'Hi! hi!' said Green, soothingly. 'Pagy! Pagy! come, now, be quiet! — will you?'

"'Let me out!' cried Hydrarchos, in great alarm. The huge bird was polking up to him. 'Let me out, I say!'

"'I never knew it to act so before,' said Green, fumbling at the lock.

"A whirr, a rush, a whizzing of the wings, and the bird was down on the doctor, treading on his heels, and pecking at the nape of his neck.

"'Pagy! Pagy!" supplicated the owner.

"But the angry bird would not listen to reason, and Sap received a thump on the head for his pains. And now both rushed for the opening door, stumbling and falling prostrate in their eagerness to escape. The monster bird danced a moment on their prostrate bodies, and then darted forth from its late prison-house.

"It rushed through a couple of grape-houses, carrying destruction in its progress. It scoured through the flower-beds, ruining the bright parterres. Mrs. Green, who was walking in the garden with her child, saw the horrid apparition, and stood paralyzed with terror. In an instant she was thrown down and trampled under foot, shrieking and clasping her infant in her arms.

"Mr. Green beheld this last atrocity, and his conjugal affection overcame his love of birds. He caught up his fowling-piece and fired at the ungrateful monster; the shot ripped up some of its tail-feathers, but failed to inflict a mortal wound,—nothing short of a field-piece could produce an impression on that living mass. Away sped the fowl to the railroad-track, down which it rushed with headlong speed. But its career was brief; an express train, coming up in an opposite direction, struck it full in front, and rushed on, scattering feathers, wings and drum-sticks, wildly in the air.

"'Tell me, doctor,' gasped Green, 'what do you think of my Great Pagoda?'

"'Great Pagoda!' said the professor, in indignant disdain. 'That was a Struthio,—Greek, *Strothous*,—in other words, an ostrich. If you had n't belonged to the genus *Asinus*, you 'd have known that, without asking me. Good-morning, Mr. Green.'

"'Where is the monster?' cried Mrs. Green. 'I believe the poor child is killed. O, Sap, I did n't expect this of you!'

"'Be quiet, my dear,' said Green; 'it was only an experiment.'

"'An experiment, Mr. Green!' retorted the lady, sharply; 'your wife and child nearly killed, and you call it an experiment! Nurturing ostriches to devour your offspring! I wonder you don't take to raising elephants.'

"'No danger of that, Maria,' replied her husband, meekly. 'I have "seen the elephant." And to-morrow I shall send my entire stock to the auction-room,— Shanghaes, Chittagongs, Brahma Pootras, Cochins, Warhens and Warhoos. They 're nice birds, great layers, small eaters, but they — *don't pay.*'"

Mr. Green was cured, of course; and though his anticipations were great, yet he had his predecessors and his successors in the hen traffic, who were almost as sanguine as he, and who not only "paid through the nose" for their experience, but who came off, in the end, really, with quite as little success. Mr. Green was but one of many. Mr. Green was one of "the people."

It will be remembered that my correspondents allude to the fowls they "*see in the noospapers.*"

I had seen these birds, in the same way, before *they* did. And a London dealer wrote me that he could send me a lot of Egleton's "famous" stock, "which took the three first premiums at a metropolitan show, and two descendants of which, at the close of the late exhibition, were sold *at auction* for forty-eight guineas ($262)."

I immediately sent out for a few of these monsters. They were described to me as being of enormous size, and *feathered upon the legs;* and I was now somewhat surprised to note that several of the English societies decided that the *true* "Cochin-China" fowl (as *they* term this variety) come only with feathered legs. The very stock above alluded to, however, came direct from the city of *Shanghae;* and duplicate birds of the same blood were delineated in the *London Illustrated News.* The metropolitan associations required that all Cochin-China fowls put in competition for premiums *must* be feathered-legged. This was a new decision, as it is well known that every importation of domestic fowl yet brought out from China direct come more or less *clean*-legged; and that fully one half of their progeny are so, with the most careful breeding, both in England and in this country. This was immaterial, however; and I repeated the story to my correspondents in good faith, and sent them copies of the portraits of these new, "extraordinary," "splendid" and "astonishing" hens, precisely as their history and pictures came to *me.* The result can be fancied. Here is the "original" portrait of

ONE OF 'EM.

This was the kind of thing that "took down" the outsiders. Orders for this strain of pure blood poured in upon me, and I supplied them. I trust the purchasers were always satisfied. In *my* case, it might answer; but I would not recommend the practice *generally* of purchasing chickens out of the newspapers. Such a portrait as the above *might* chance to be a little fancif ıl; or, *perhaps*, it might be a trifling exaggeration, you see. Yet this was the breed that were always "put in the *newspapers*." You very rarely found them in your coops, though!

CHAPTER XXVI.

"POLICY THE BEST HONESTY."

THIS reversion of the old saying that "honesty's the best policy" seemed to have finally attained among many hen-men, and the ambition to dispose of their now large surplus stock, at the best possible prices, had become very general, while the means to accomplish it came to be immaterial, so that they got rid of their fancy poultry at fancy figures.

Nothing that could be said against me and my stock was neglected, or omitted to be said. But, as long as fowls would sell at all, I had my full share of the trade, notwithstanding this. The following veritable letter, received from a noted "breeder," in 1853, will explain itself; and it exhibits the disposition of more than *one* huckster still left around us. It will be observed that this gentleman called me his "friend"!

"FRIEND B——: What has become of all the trade? I have n't sold twenty dollars' worth of chickens, in a month! I've now got over three hundred of these curses on hand —

and they 're eating me up, alive. What 'll we do with them? Do you want them? Will you buy them — *any-how*? And give what you like for them.

"They are a better lot than you ever owned,— every-body says so,— Greys, Cochins (*pure*) and Shanghaes. D—n the business! I 'm sick of it. My fowls and fixin's cost me over twelve hundred dollars. What do you think of an auction? Has the bottom fallen out, entirely? Could I get back two or three dollars apiece for this lot, do you think, at public sale?

"B—— is stuck with about five hundred of the gorman-disers. I 'm glad of it — glad — *glad!* An't you? He always lammed you, as well as me; and though I think *you* can swinge the green 'uns as cutely as 'most any of 'em, *he* has been an eye-sore for three years that ought to be put down. He got his stock of you, he says,— but (no offence to you, friend B——), it an't worth a cuss. All of it 's sick and lousy, and he shan't sell no more fowls, if I can help it.

"Have you seen W——'s stock, lately? Is n't *he* a beauty! I told him, last week, he 'd ought to be ashamed of himself ever to gone into this trade, at all. He 's well enough off, without stealing the bread out of the mouths of them that 's a long way honester than he ever was. I 'll have a lick at *him*, yet.

"Come and see my stock,— and buy it. I don't want

it. I must give it up. I'm too busy about something else. Come — will you? I don't say anything against your fowls, outside; but you know, as well as I do, that you have n't got the *real thing*. Bennett says you have n't, and everybody else says so. As to your 'importations,' you never had a fowl that was imported from any further off than Cape Cod, and you know it! But that is neither here nor there. *I* don't care a fig how much you gouge 'em. All I want is to get rid of *mine*. If you don't buy them, I shall sell them,— somehow,— or give them away, sure. They shan't eat me up, nohow.

"They don't eat nothing — these fowls don't! O, what an infernal humbug this is! I never got much out of it, though. I tell everybody what all the rest of you do,— of course. But *I* had rather keep the same number of Suffolk pigs, anyhow, so far as that 's concerned. I an't afraid of your showing this letter to nobody — ha! ha! So I don't mark it 'private.' But of all the owdacious humbugs that ever this country saw, *this* thing is the steepest,— and you know it!

"Write me and say what you 'll give me for my lot. I won't peach on you. You can buy 'em on your own terms. I want to get out of it. And you may say just what you 've a mind about 'em. I 'll back you, of course. Could n't you take them, and get up another fresh guy on a 'new importation'? That 's it. Come, now, friend

B——, help me out. And answer immediately. All I want is to get out of it, and catch *me* there again if you can! Yours, &c.,

" —— ——.

"P. S. If you don't buy them, I shall kill the brutes, and send 'em to market; though they are too *poor* for that, I think."

This complimentary epistle from a brother-fancier was rather cool, but it did n't equal the following. I had more than one of this sort, too,— of which I had no occasion, for the time being, to take the slightest notice, for I had "other fish to fry," decidedly!

"MR. BURNHAM.—SIR: How is it that you have the impudence to try to palm off on the public those fowls of yours for genuine '*imported* ones,' when it is known that you bought them all of me, and A——, and B——? How can you sleep nights? Don't you feel a squirming in your conscience? Or is it made of ingy-rubber, or gutter-perchy? You have made hundreds, and I don't know but thousands of dollars, by your impudence and bare-faced deceit. They are *not* genuine fowls. I say this *bolely*. I wish there was a noospaper that would show the inderpendence to print an article that I could rite for it, on this subject of poletry. If I would n't make you stare, and shet your eyes up, too, then I aint no judge of swindling!

"Why don't you act like a man? *Carnt* you? Havn't you got the pluck to own up that other people have done for you what you never had the gumption to do for yourself? Why don't you act fair,—and tell where the genuine fowls can be got, and of who? You 're a doing the poultry business more hurt than all the rest of the men in the country is doing, or ever did, or ever will, sir.

"I don't mind a man's being sharp, and looking out for himself. *I* do that. But I carn't humbug people as you are doing,—and I won't, neither. You 're sticking it into the people nicely,—don't you think you are? And they *believe* it, too! The people believes what you tell them, and sucks it all down, and wants more of it. And you keep a giving it to them, too! How long do you suppose such infamous things as these can last? I hope this letter will do you good. I havn't no ends to answer. I keep but a few fowls, and I have never charged over twenty-five dollars a pair for the best of them,—as you know. *You* get fifty or a hundred dollars a pair. So the noospapers say, but I believe you lie when they say so. You carn't come this over *me!* You don't pull none of that wool over *my* eyes! No, sir!

"If you want to get an honest living,—get it! I don't say nothin against that; you 've a rite to. But don't cheat the people out of their eye-teeth, by telling these stories that you carn't prove.* You 've no right to. You sell fowls, by this means, but you don't get no clear conscience by it. It 's wrong, Mr. Burnum, and you know it. While you do this, nobody can sell no fowls except *you.* Give other people a chance, say I. I would n't do this, nohow, to sell my fowls at your expense; and I go for having everybody do unto others as *I* would do to *them.* This is moral and Christian-like, and you 'd better adopt it. That 's my advice, and I don't charge nothing for it. So, no more at present — from Your, resp'y,

"—— —— ——."

These missives never disturbed me. Why should they? These very men would have sold, from that very stock,— *had* done so, repeatedly, before,— whatever a buyer sought to purchase. I never knew either of them to permit the chance of a sale to pass by him, on account of the *variety* of bird sought! They invariably possessed whatever was wanted. With them, "*policy* was the best *honesty*." I did not complain. I was a "hen-man," but no Mentor.

* I never found, in my limited experience in this business, any particular necessity for attempting to prove anything. "The people" wanted FOWLS — not *proofs!*

CHAPTER XXVII.

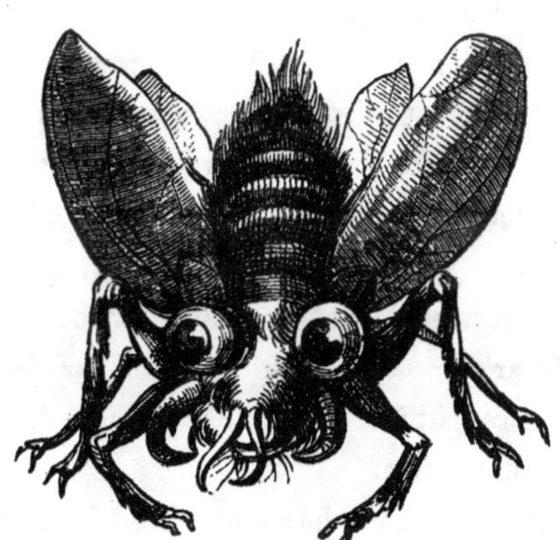

A GENUINE HUMBUG.

It was now getting pretty clear to the vision of most of the initiated that the hen fever was in the midst of its height. Buyers with long purses were about, but they were not sc ravenous as formerly. They talked knowingly and cautiously, and chose their fowls with more care than formerly; but still a great many samples were being circulated, and at very handsomely remunerating prices.

A gentlemanly-*looking* man called upon me, one day,

about this time, in Boston, and introduced himself, in his own felicitous manner, something in this wise:

"How are you? Mr. Burnum, I suppose. My name is T——. I 'm from Phil'delphy."

"Happy to see you, Mr. T——," I replied. "Take a seat, sir?"

"I want to *look* at your fowls, Burnum," he continued, in a rather bluff manner. "I know what poultry is, I *think*. I've been at it, now, over thirty year; and I 'd oughter know what fowls is. You 're a humbug, Burnum! There 's no doubt about *that;* and you 're all a set of bums, together — you hen-men! I have n't got the fever. I 'm never disturbed by no such stupid nonsense. These China fowls are an old story with *me*. I had 'em twenty years ago,— brought into Phil'delphy straight from Shanghae by a friend of mine."

[This gentleman had forgotten, or did n't know (or thought *I* did n't), that the port of Shanghae had been open to communication with this country only a dozen years or less; and so I permitted him to proceed in his remarks without offering any opposition to his assumption.]

"These big fowls never lay no eggs, Burnum. You know it as well as anybody. *Do* they?"

"None to hurt," I answered.

"No, no — I reck'n not," continued my visitor. "*I* know 'em, like a book. Can't fool *me* with them. They an't worth a curse to nobody. I'll go out and *see* yours, though, 'cause you 're a good deal fairer than I expected to find you. I thought you 'd try to hum *me*, same as I s'pose you do the rest."

"O, no!" I replied, meekly. "When I meet with gentlemen who are posted up, as *you* are, sir, I conceive it to be useless to attempt to urge them to possess themselves of this stock; because I am always satisfied, at first sight, what my customer is. And I govern myself accordingly. I will take you out to my place, directly. My carriage is in town, and we 'll ride out together. You can see it,— but you say you don't want to purchase any?"

"No, no — that 's not my object, at all. Still, I like to look at the humbugs, any way."

I was as well satisfied that this man knew very little of what he thus boldly talked of, as I also was that he had come all the way from Philadelphia *purposely* to buy some Chinese fowls. But I gave him no hint of this suspicion; and we arrived, an hour afterwards, at my residence in Melrose.

He examined my fowls carefully; went through all the coops and houses, and finally we entered the "green-house" where the *selected* animals were kept. As soon as he saw these birds, *I* saw that he was "a goner."

He denounced the whole race as he passed along; but when we entered this well-appointed place, he stopped.

These were very respectable, and he would n't mind having a few of *these*, he said.

"What do you get for such as these?" he inquired.

"Twenty-five dollars each," I replied, "when I sell them. But they 're all alike. *You* know it as well as I do. They 're worth no such money. These fowls are well-grown, and are in good condition; but five or six shillings each is their full *real* value. Still, you know when 'the children cry for them,' why, we get a little more for them."

"Yes; but twenty-five dollars is a thundering hum, any-how, Burnum! I can't go *that!* You must n't think of getting no such price as that out of *me*, you see; 'cause you know that *I* know what all this bosh means. I 'd like that cock and those three big hens," he added, pointing to four of my "best" birds. "That is," he continued, "if I could have them at anything like a fair rate."

"My dear sir," I responded; "*you* don't want any such hum as this imposed upon *you*. You know, evidently, what all this kind of thing signifies. But, at the same time, you see I can get this price, and do get it every day in the week, out of the 'flats' that you have been speaking of. I don't sell any of these things to gentlemen, who know, as *you* do, what they are, you see."

"Yes, yes!" continued the stranger; "I know; I see. I comprehend you, exactly — precisely. But I should like them four fowls. What 's the *lowest* price you 'll name for them ? "

" I never have but one price, sir," I replied. " *These* fowls I keep here for show-birds. They are my ' sign,' you perceive — my models. The younger stock, that you have seen outside, are bred *from* these ; and thus I am enabled to show gentlemen, when they come here, what the others will be " — (*perhaps*, I might have added ; but I did n't).

This gentleman remained half an hour at my house, and we talked the whole subject over, at our leisure. I agreed with him in every proposition that he advanced, and he finally left me with the assurance that I had been traduced villanously. He really expected to meet with a regular sharper when he encountered me ; but he was satisfied, if there was a gentleman and an honest poultry-breeder in New England, *I* was that fortunate individual !

I did not dispute even this assurance on his part. And when he left, *I* had one hundred dollars of his money, and *he* took away with him four of my " splendid " pure-bred Grey Shanghaes, which I sent to the cars with him when he bade me good-day.

This was but a single sample of the *real* humbugs that presented themselves to us, from time to time, *all* of whom were certain to inform us that they were " thoroughly acquainted " with the entire details of the business ; all of whom had been through the routine, and "knew every rope in the ship;" none of whom were affected with the "fever" (so they always declared), and not one of whom believed, while they were thus striving to pull wool over the eyes of others, that they were all the time being "shaken down" without mercy!

This was the very class of men who, in the later days of the malady, assisted most to keep up the delusion, and to aid in carrying on the hum of the trade. To be sure, the keepers of agricultural warehouses talked, and told big stories to their poor customers, who would buy eggs and chickens of them, for a while, at round prices; true, most of the agricultural papers strove from week to week to keep up the deceit, after the editors or proprietors found their yards over-stocked with this species of property, for which they had originally paid me (or somebody else) roundly, and which they "could n't afford to lose," though they *knew* it to be valueless! True, the hen-men themselves kept their advertising and the big stories of their success constantly before "the people," whom they gulled from day to day. But no portion of the community did more to "help the cause along" than did this self-sufficient, learned, know-nothing, thin-skinned class of "wise-acres," who never chanced to make much more than a considerable out of the writer of this paragraph — I *think!*

Among this well-informed (?) set of men there was

a "John Bull" who was connected in some way with a Boston weekly, which was nominally called an agricultural sheet, but which for several years was filled with articles upon the subject of "the equality of the sexes."

His name was Pudder, or Pucker, or Padder, as nearly as I remember. From the commencement of this fever he was sorely affected, and his articles upon the merits of the different breeds of fowls he raised were very learned and instructive! He sold eggs for three, four, or five dollars a dozen, for a few weeks; but, as they did n't hatch, his game was soon blocked. Still, he stuck to this hum with the obstinacy of a "bluenose;" and his readers were indebted to his advice for possessing themselves of the most worthless mass of trash (in the shape of poultry) that ever cursed the premises of amateur. His lauded "Plymouth Rocks," his "Fawn-colored Dorkings," his "Italians," his "Drab Shanghaes," etc., sold, however; and the poor devils who read the paper, and who purchased this stuff, lived (like a good many others) to realize, to their hearts' content, after paying this fellow for being thus humbugged, the truth of the old adage that "the fool and his money is soon parted."

Still, Podder was useful — in his way — in the hen-trade. The operations of such ignorant and wilful hucksters had the effect of opening the eyes of those who desired to obtain *good* stock, and who were willing to pay for it. And after they had been thus fleeced, they became cautious, and procured their poultry only of "honorable" and responsible breeders (like myself), who imported and bred nothing but known *pure* stock.

As late as in January, 1855, a western agricultural sheet alludes to the flaming advertisement of an old hand in this traffic, and says: "It is known to all who know anything about poultry that Mr. G—— has been an amateur breeder for about forty years, and is undoubtedly better 'posted,' in reference to domestic and fancy fowls, than any other man in America; and, beside this, he is an honest man, and has no 'axe to grind.' He has raised fowls, heretofore, *solely for his own amusement;* but *now* he proposes to accommodate the public by disposing of some of them."

This man is my "fat friend" in Connecticut,— who has bred and bought and sold as much *trash*, in the past ten years, as the best (or the worst) of us. Friend Brown, we could tell you a story worth two of yours, on this point! But — we forbear.

CHAPTER XXVIII.

BARNUM IN THE FIELD.

THE prince of showmen was suddenly developed as a "hen-man"! Mr. Barnum was seized, one morning, with violent spasms, and, upon finding himself safely within the friendly shelter of "Iranistan," his physicians were duly consulted, who examined his case critically, and reported that the disease lay chiefly in the head of their patient — who, it was subsequently ascertained, was suffering from a severe attack of hen fever.

Such was the violence of the demonstrations in this gentleman's case, however, and so fearful were the indications with him, even during the incipient stages of the affection, that his friends feared that Phineas T. had really contracted his "never-get-over." But, upon being informed (as I was, soon afterwards) of this case, and questioned as to his probable eventual recovery, I unhesitatingly gave it as my opinion that his friends might rest assured the humbug that could kill *him* was yet to be discovered; and that, so far as he was personally concerned, I entertained no sort of doubt that "he would feel much better when it was done aching." (A prediction which, I have no question, has been accurately fulfilled, ere this.)

The man who could succeed, as he had, with no-haired horses, gutta-percha mermaids, fat babies, etc., and who had gone into and out of fire-annihilators, prepared mastodons, illustrated newspapers, copper mines, defunct crystal palaces, and the like, unscathed, would scarcely be jeopardized by an attack of the prevailing malady of the day, however violently it might exhibit itself in his case. And so there was hope for Phineas, though his symptoms were really alarming.

My friend took the very best possible means for alleviating the virulence of his attack; and, looking about him for the largest-*sized* humbug known in the trade, he alighted upon a two-hundred-and-forty-pound Connecticut joker, who quickly offered to inform him how he could find relief.

"How shall I do it, John?" exclaimed Phineas, as his fat friend made his appearance.

"Heesiest thing in life," responded John; "hall you 'ave to do is to put yer 'and in yer pocket."

"*So?*" said Phineas, putting his fist gently out of sight.

"No—you aren't deep enough down yet," replied John. "Go down deeper. That's better,—that 'll do."

"How much 'll it cost?" queried Phineas.

"Carn't say," responded John. "You 're pooty bad. There 's nuth'n' in *this* country that 'll cure you. Hi 'll go hout to Hingland, if you say so, and hi can git somethin' there that 'll 'elp you. It ar'n't to be 'ad in Ameriky, though."

"Sho!" exclaimed Barnum; "you don't say so! Do you think, John, that we could find something in England that would knock 'em, here?"

"Nothing else," replied John. "*Hi* know where they keep 'em." (John was raised in Great Britain.)

"But, John," persisted Phineas, "there 's *Burnham*, you know, of Boston. They say *he* has the best poultry in the world; and I 've no doubt of it, between you and I."

"Fudge!" exclaimed John; "Burn'am 's a very clever fellow, hi 've no manner o' doubt, and hi won't say nuth'n' ag'inst 'im; but 'ee 's the wust 'umbug you *ever* see, since you 'ad breath. 'Ee don't know the dif'rence 'tween a Shanghi and a Cochin-Chiny — an' never did. 'Ee 's a *hum*, 'is Burn'am. Don't go near '*im*, unless you want the skin shaved hoff o' yer knuckles, clean."

"Well, John," said the show-man, "something must be done. I 've got the fever, bad, I 'm afraid, as you suggest; and it must be fed. What can you do for me?"

John thought the matter over, and it was finally agreed, as there were no good fowls in America (according to John's notions), that *he* should be deputized by Phineas to proceed to "Hingland," and procure some genuine (that is, *pure*) stock, for the coops at Iranistan, at the liberal show-man's expense! A capital recipe, this, for Barnum's disease, as well as for John's own benefit.

But Phineas is n't taken down easy, though they do occasionally "fetch him." And so he hesitated. He thought the matter over a while, and finally said to his friend, one day,

"John, I 've got it!"

"'Ave you?" says John.

"Yes, I 've got it. You know I 've something in my head besides grey hairs, John."

"Hi 've no manner o' doubt o' *that*," replied John.

"Well, I have thought this thing over, and I have determined to see, first, what there is in America, before I send you out to Europe."

"It 'll take you a long time to do that," said John, "and you 'd 'ave to travel a great w'ile to see all the poultry we 've 'ere."

"I won't travel at all," said Phineas.

"No? As 'ow, then?" inquired John.

"I 'll get up a show — a poultry exhibition — on a grand scale, and it shall come off at my Museum, at New York. Everybody 'll come, of course; and we can see what there is, buy what I want from the best of 'em, and make our selections as we may fancy; you shall go out afterwards to England, and obtain for me what I can't get here, you see."

"Capital! — hexellent!" responded John.

"And I'll call it the — the — *what?*" said Barnum, stopping for an appropriate title to this anticipated exhibition.

"I donno," said John, puzzled.

"Well — then — the *National Show*," continued Phineas. "How'll that do? The first exhibition of the 'National Poultry Society.' I think that's good. You see that includes all quarters of the country; and we shall know no north, no south, no east, no west! A quarter admission — Museum included — capital!"

"Yes—just *the* thing!" chimed in his friend. And shortly afterwards advertisements and circulars found their way into the hands of all the hen-men in the country, who were thus invited to visit New York, in February, 1854, to contribute to the grand show of the "National Poultry Society," of which P. T. Barnum, Esq., was *President*.

A long string of names was attached to this call, and the list of "Managers" embraced one or more representatives from every State in the Union —my own humble name appearing among the Vice-presidents for Massachusetts.

The whole thing was clearly one of Barnum's *dodges* to fill his Museum for a few days; and probably not a single individual except himself had any knowledge of the formation or existence of any such *society* as this, of which he thus nominally appeared to be the presiding officer. At any rate, after diligent inquiry, I could never ascertain that anybody knew anything about any such an association, except himself.

However, this was a matter of no sort of consequence, of course. The Fitchburg Dépôt Show, in Boston, was a similar affair; and I now joined in this exhibition without asking unnecessary questions,—because I saw that there was fun ahead, and that *I* could make an honest penny out of it, whether Barnum did or not.

Every one now put his best foot foremost; and, as this fair approached, Shanghaes were converted into Cochin-Chinas (by the knowing ones), by the removal of the feathers from the legs; the mongrels were made feathered-legged Bother'ems, by the free use of gum-tragacanth and down; the long-tailed fowls were deprived of all superfluous plumes, through the aid of the pincers; and what this last process did not satisfactorily effect, the application of the *shears* completed (see engraving!); until, at last, the unlucky bipeds, whom nature had originally supplied with decent caudal appendages, were reduced to that requisite state of brevity, astern, which the *mode* or the taste of the day demanded. And, at length, all was ready for the great "National Show" in New York city.

As it turned out, the whole thing (though an utter *sham*

PREPARING FOR THE FOWL SHOW. — (See page 195.)

as regarded its being a *society* matter) proved to have been well conceived, and, from beginning to end, was admirably well carried out. Mr. Barnum did his part most creditably at this first show in New York, and the experiment was eminently successful.

The birds were afforded excellent care, and an immense quantity of good specimens found their way to the Museum at the appointed time. For a week, notwithstanding the very dull weather, the great rooms of the American Museum on Broadway were thronged with visitors; and Barnum was in high glee at the entire success of his undertaking.

Not content with one week's show of the fowls, Barnum proposed that it should be continued for six days longer; and the crowd continued to visit this exhibition for another week, and to pour in with their friends, their wives, their children, and their quarters, to the great edification and satisfaction of the proprietor of the show, and the "President" of the "National Poultry Society."

I was there, with a goodly quantity of my "rare" and "unexceptionable" and "pure-bred" fowls, which were greatly admired by the thousands of lookers-on, who flocked to this extraordinary exhibition. It was really astonishing (to *me*, at least) what very fine birds I had at this show.

And, "may be," fowls did n't *sell* there! If I remember rightly, "the people" were round, on that occasion. And so was *I!*

CHAPTER XXIX.

FIRST "NATIONAL" POULTRY-SHOW IN NEW YORK.

WHETHER it was because Barnum had taken this enterprise in hand, whether it was because it was known that my "superior" stock was to be seen at the Museum, or whether it was because the intrepid "Fanny Fern" had promised to visit the show, I cannot say; but one thing was certain,—such a gathering of "the people" was seldom witnessed, even in busy, driving, sight-seeing New York, as that which crowded the great rooms of Barnum's establishment on the occasion of the first exhibition of the so-called "National Poultry Society."

"All the world" was there, with his wife and babies, and nieces and nephews. The belle and the beau, the merchant and the mechanic, the lawyer and the parson, the rich and the poor, old and young, grave and gay,—all were in attendance upon this extraordinary display of cockadoodledom; and Barnum—the indefatigable, the enterprising, the determined, the incomparable Barnum—was in his glory, as the quarters were piled up at the counter of the ticket-office, and "the people" wedged their way up the crowded stairs and aisles of his Museum.

The great show-man was as busy as His Satanic Majesty is vulgarly supposed to be in a snow-storm! Now here, now there; up stairs, down stairs; in the halls, in the lobbies; busy with John, button-holing the "committees," from morning till night. All smiles, all good-nature, all exertion to please the throngs of visitors who constantly jammed their way about the building. And, to say that everything about this undertaking (so far as he was personally concerned) was not managed with tact and good judgment, as well as complete propriety and liberality, would be to state what was untrue. Mr. Barnum rarely does anything by halves; and to him, in this instance, belongs the credit of getting up, and carrying through successfully, the very best show of poultry ever seen in America,—beyond all comparison.

In due season I selected from my then somewhat reduced stock sixty specimens of the Shanghae tribe of fowls, which, with some twenty samples of choice Madagascar Rabbits, I forwarded (in charge of my own agent) to this long-talked-of show.

The person whom I employed to look after my stock — (for I had long since got to be "a gentleman," and could n't attend to such trifling matters, personally) — the man who went with it to this exhibition was thoroughly posted up in his "profession," and knew a hawk from a handsaw, as well as a Shanghae from a Cochin-China. And when he started for New York with my contributions, I enjoined it upon him to bear in mind, under *all* circumstances, that the gentleman he represented had the only *pure*-bred poultry in America, any way. To which he replied, briefly,

"Is *that* all? I knew that before."

I said, "John, you 're a brick. A faced-brick. A *hard*-faced-brick. You 'll *do*."

John winked, and left me, with the understanding that, as soon as he should have time to look around the show, he would telegraph me at Boston what the prospect was, comparatively. I felt quite *sure* that my fowls would take all the premiums, for they always had done so before; and my "pure-bred" stock grew better and better every year!

I did not go to the show for a day or two after my agent left; and, on the morning succeeding the opening, I received from him the following brief but expressive telegraphic dispatch:

"G. P. Burnham, Boston.

"Arrived safe; thought we 'd got 'em, *sure*. We have — *over the left*. You are nowhar! B."

Here was a precious fix, to be sure! For five years, I had carried away the palm at every exhibition where my "splendid" and deservedly "unrivalled" samples had been put in competition with the stock of others. And now, at the first great *national* exhibition, where everybody would of course be present (and where the first cages that would be looked for, or looked into, must be those of Mr. Burnham, the breeder of the only original "pure"-blooded poultry in the country), according to my agent's dispatch I was *no*whar!

This dispatch reached me at noon, and on the following morning I was in New York. I looked about the several apartments in the Museum, and satisfied myself who had the best fowls there, very quickly. As it happened, they were not inside of *my* cages, by a long mark!

Yet "the people" crowded around my showy coops, for which my agent had secured an advantageous position, and in displaying them (if I remember aright) he lost no opportunity in saying just *enough* (and no more) to the throng who passed and admired their beautiful proportions, their great size, and splendid colors. There were not a few choice birds scattered about the rooms,—under the benches, or in the far-off corners,—which my eye fell upon, which my agent subsequently purchased at very modest prices, and which found their way, somehow, into my coops.

"The people" now stared with more earnestness than ever. By the evening of the second day, my "pure-bred" stock *did* look remarkably well! And when the "committee" came round, at last, I found myself the recipient of several of the leading premiums, for my "magnificent," "superb" and "extraordinary" contributions, again. And now commenced the fun, once more, in earnest.

Everything that I sent to New York was quickly bought up at enormous prices. Fifty, eighty, a hundred, a hundred and twenty-five dollars per trio, was willingly paid my agent for the rare and incomparable fowls I exhibited there. "The people" were literally mad on the subject; and I had n't half enough to supply my customers with, at figures that astonished even *my* ideas of prices,—which, by the way, were not easily disturbed!

During this exhibition, Mr. Barnum announced that a "conversational" gathering would be held, one day, in the lecture-room of his Museum; whither the throng were invited to epair, at last, to talk over matters pertaining to the weIfare of the trade generally, and the hen-humbug more particularly.

A rush was directly made for this hall, which was quickly filled up by the multitude, who now stood or sat, with gaping mouths and staring eyes, in readiness to be further bamboozled by the managers of this *National* "Society," who duly paraded themselves upon the platform, and commenced to show themselves up for the edification of the uninitiated, and to the great amusement of those who had "been there" before them.

Mr. Barnum presided, but with that grace and modesty and extreme diffidence for which he is so noted. The enthusiasm of the occasion soon reached concert-pitch, however, and everybody on the stage, in the parquette, and around the gallery, desired to relieve themselves of the pent-up patriotism that rioted in their bosoms; and all desired to be heard at the same time.

Cries of "Barnum! Barnum!" "Where's Bennett?" "Speech from Burnham!" "Down in front!" "Give 'em a chance!" "Hear the president! — there he is!" "Hurra for the Bother'ems!" &c. &c., rang from the lungs of the crowd. And finally order was restored, and Mr. Barnum approached the front of the stage, to deliver himself of "feelings that could be fancied, not described," amid the cheers and shouts of that crazy multitude.

CHAPTER XXX.

BARNUM'S INNATE DIFFIDENCE.

As soon as the vociferous cheering had subsided, Mr. Barnum reached the foot-lights, and smiled beneficently upon the crowd before him.

"LADIES AND GENTLEMEN," said the show-man, modestly, "unaccustomed as I am to public speaking, you will pardon me, *imprimis*, for hinting at the extreme diffidence with which I now rise to address you; and I am sure that, notwithstanding the commendable zeal that now animates this enlightened audience, you will sympathize with me in the midst of the embarrassments under which you must readily perceive *I* am laboring, and extend to the speaker your lenity (all unused, as you are aware he is, to this sort of scene), while he ventures upon a few very brief remarks on the interesting and laudably-exciting topic that has brought us together here, on this happy occasion."

This modest appeal brought down the house, of course; and the bashful Mr. B., after clearing his throat, was requested by the crowd to "Go on, Barnum! Proceed—put 'er through!"

"The hen fever," continued Mr. B——, "is but just *begun* to be realized, ladies and gentlemen, among us." (Barnum had been attacked by the malady only a few weeks previously, and had n't "heard from the back counties" then!) "This first exhibition of the National Poultry Society, my friends, is ample evidence in support of this statement. Was there ever such a show seen, or heard of, ladies and gentlemen, as this which you are now the witnesses of? Never! Yet, I repeat it, this is but the commencement. The enthusiasm which has attended upon this exhibition, the feelings that have been stirred up by this before unheard-of display, the people of every grade in society that come forward here in its support, the zeal which animates the bosoms of the thousands upon thousands who have attended it, and the names of the men connected with its origin and present patronage, afford ample evidence in support of my assertions, that the fire has but just begun — just *begun* to burn, fairly, ladies and gentlemen!" ("That 's a fact," was the ready response of a young gentleman who had just paid my agent over three hundred dollars for a few samples of my "choice" chickens; the first he ever owned!)

"*I* want to say a few words," remarked a stranger, under the gallery, at this point. But he was requested by the chairman to "*hold in!*" until Mr. Barnum concluded. After considerable urging, this anxious man was prevailed upon to sit down; though he was evidently "full to bursting," with his enthusiastic emotions.

"We have a good deal to learn yet, gentlemen," continued Barnum (and that was truthful, at any rate!) "We have much to learn; but we know enough to spur us on to acquire more. More knowledge, more experience, more fowls. We have n't enough — we don't know enough, yet. I am greatly rejoiced at the prospects, to-day, and with the entire success of this enterprise, here!" (And well he might be.) "I have freely given my time and humble talents to its consummation, and we have triumphed! We, *the people*, the men who have the heart and the pluck to undertake and carry through this sort of thing. There 's no hum in *this*, gentlemen! None, whatever. How *can* there be? We see this thing before our very eyes. It is a tangible, living, breathing, walking, crowing" (and he might have added *eating!*) "reality, ladies and gentlemen. There can be no humbug in anything of this sort; because we can take hold upon it, handle it, view it with our eyes open. A humbug is but an unexplained or half-concluded *fact*. This is a self-evident, clearly-defined fact —

'A thing that *is* — and to be blessed!'

And when you, or I, can take a crower in our hands that will weigh twelve or fourteen or fifteen pounds,— when we can see and feel him,— can there, by any possibility, be humbug in it?"

"No — no — no!" shouted the crowd; the ladies kindly joining in the decisive negative given to this forcible appeal.

"Then, I repeat it, we are but just in the beginning of the commencement of this new and promising era. The fire has just begun to burn, and to illumine the world; and, as I said before (or intended to say), it is not to be subdued! It is a mighty conflagration, which assails everybody at this moment, and is now enveloping all classes of the community, from the highest to the lowest! This land is in a blaze! In a threatening, exciting, violent, whirling, astounding blaze, gentlemen — and no opposition or invention can put it out!" ("Fetch on your fire-'nihilators, then!" shouted a vicious wag, from the gallery.)

"We don't want to put it out," continued Mr. Barnum, growing warmer as the fire of his zeal in this cause continued to glow within him; "we have no wish to put it out. Let it burn! Let it come! Let it conflagrate! We love it — *you* love it — *I* love it — it's one of the things we admire to think of, and speak of, and read of, and pay for, and help to keep alive here, and everywhere, and elsewhere! Our country is big enough; we have millions of broad acres, miles on miles of fertile fields, and cords of maize and grain that cannot be used or disposed of, unless it be devoted to the uses and benefits of these beautiful birds, sometimes so cavalierly spoken of by their enemies, but the value of which *I* know, and most of *you*, gentlemen, know how to appreciate!" (Applause, and cries of "Go it, old hoss! You 'll be a capital customer for some of the hen-men to pick up! Go it, Barnum!")

"I did not rise, gentlemen," continued the speaker, "with any idea of telling you anything new. I am but an humble coadjutor with you in this pleasing and innocent undertaking. I can see, as you can, also, the importance of this subject" (he did n't say *what* "subject"), "and I trust that we may go on, and increase, and multiply domestic fowls and customers, in a ratio commensurate with the rapidly increasing throbs of the public pulse — which is now beating only at 2.40, and which must soon reach a 2.10 pace, if nothing breaks!" ("Hurra! Hurra!" yelled the boys; "that 's a good 'un!") And the President sat down, blushing, amid the uproarious applause that followed his remarks.

As soon as order was comparatively restored, other gentlemen, whom the President introduced as "honorable," and "talented," and "professional," and "influential," took the rostrum, and "followed suit" upon Barnum's lead.

A vote of thanks was finally passed to Mr. Barnum for his services, and the *sacrifices* he had made in behalf of the "Society;" another to the "orator" of the day (whose name I have now forgotten), formerly a member of Congress, I believe; another similar vote to the Secretary, to whom, also, a plated jug was subsequently presented; a vote to Mr. Burnham, of Boston, for his speech and his "magnificent" contributions of *pure*-bred stock; a vote condemning everybody who had or should thenceforward nickname fowls; a vote of condolence and sympathy with John Giles, because none of his *pure* Black Spanish fowls were in the exhibition; a vote to Porter, of the New York *Spirit of the Times*, for his disinterested notices of the show; another to Greeley, of the *Tribune*, who had n't time to visit it; another to pay the bills of the "Committees" at the Astor House (*minus* the champagne charges!); another to Dr. Bennett, for not being present at this show; another endorsing the claims of patent pill-venders and cross-grained bee-hive makers; another to Frank Pierce, for the allusions in his inaugural to the "march of progress" in our land, which of course included Shanghae-ism; another to Caleb Cushing (an honorary member), who was lauded as the most thoroughly graceless humbug known to the "national" society; another endorsing the collector and postmaster of Boston as disinterested democrats; another that my "Grey Shanghaes" were evidently the only full-blooded fowls exhibited at the American Museum on this occasion; and numerous other resolves were duly "voted," of which no note was taken at the time.

While this bosh was transpiring, I sent to Boston for some fifty pairs more of my "superb" specimens of Shanghaes and Cochins, all of which were disposed of during the second week of this show, at curiously "ruinous" rates. And at the close of the exhibition my agent had taken very nearly *three thousand dollars* for the "pure" Shanghaes, and Cochins, and Greys, he had sold there for my account!

I trust that every one was as well satisfied with the results of this first exhibition of the "National Poultry Society" as I was. It is the last show *I* shall ever attend. And having invariably taken the lead, from the beginning up to this trial, I retired, content with the self-assurance that I had made all I could make out of this sort of thing, and that the field now legitimately belonged to my juniors in the profession. May success attend them!

At the close of the exhibition, my friend Barnum congratulated me.

"They tell me you 've done *well*, Burnham," said my friend, cheerfully. "I 'm glad of it. And, since you 've made it so handsomely, suppose you leave me a couple of your best Fancy Rabbits, yonder; I 'll add them to the 'Happy Family.'"

"Certainly," I replied. "With great pleasure, B——. And, since *you* have done so capitally with this show, you shall give me a quarter of your profits on the tickets sold. Here — take the rabbits!"

"A-*hem!*" said Barnum. "No — no. It 's no matter. You need n't — no — we won't say anything about it. It 's all right. You 'll do. You can run alone, I guess. *I believe I don't spell my name right!* Good-by — good-by."

I have n't seen friend Barnum since.

At this exhibition of poultry I managed to show a pair of my pure-bred Suffolk pigs, too, which did not set me back any. I took numerous orders for these animals, and I have given on page 174 what passes for a likeness of a fancy "Shanghae" fowl, such as we "read of in the newspapers," and which everybody, during the last five years, imagined he was buying, when he ordered "such," after seeing the "pictur'."

In this class of illustration, there was quite as much deceit and chicanery practised, commonly, as in any part of the general system of the humbug. The uninitiated saw the well-rounded forms of the huge fowls or hogs he sought, in his weekly agricultural journal, from time to time; and, through the same channel, he met with "portraits," represented to have had originals at some time or other, and which were said to be in the possession of this or that breeder, who "had been induced, after earnest solicitation, to part with a very few choice samples," out of such imaginary stock. With the *swine,* the thicker the ham, the smaller the feet, the shorter the nose, and the thinner the hair, the better and the *purer* blooded pig you got, for instance!

The following is a sample of this kind of guy, which has had its run in the past three years, and upon which tens of thousands of dollars have been squandered by enthusiastic admirers of these bloated bladders of lard. This is *supposed* to be a likeness of the "genuine"

SUFFOLK PIG.

The good old lady replied, when asked if she loved the Lord, "I donno much about him, but I hain't nothin' agin him!" So I affirm in reference to this hog. But one thing I may be permitted to remark in this connection; to wit, that the more pure Suffolk *pigs* there are, the less *corn* you find round. That 's all!

CHAPTER XXXI.

A SUPPRESSED SPEECH.

THE following remarks, on the occasion referred to, were neither published at the time, nor would the "Committee on Printing" admit them into the official report of the proceedings of this *national* show. For what reason, I am utterly unable to determine. These were the author's sentiments, and I give the speech a place *here*, because I have no idea of being thus "headed" by my colleagues in that enterprise. This speech was delivered by the Young 'Un "with emphasis and discretion;" but the managers suppressed it. I now submit it, in the hope that it will be duly appreciated. When called upon, I said, as modestly and as gracefully as I knew how:

"MR. PRESIDENT: *Vox populi, vox Dei!* The people assembled within the classic and well-painted walls of your American Museum call upon me for a few words of encouragement; and, while I assure you I find myself totally unprepared to speak (though my present address has been written some four weeks), I cheerfully respond to the flattering demonstration that greets me on this electrifying occasion." (Applause, and waving of hats and handkerchiefs.)

"I am but an humble disciple in this profession, Mr. President, and know very little of the deceit and chicanery that *some* persons charge others with practising in the ramifications of the hen-trade; and, although it has been said that 'what I don't know about this part of the business would n't be worth much to anybody,' yet I here solemnly disclaim any superhuman or supernatural knowledge of the tricks of this laudable and highly respectable calling." (Cries of "Good, good! You 're an injured man! Go on!")

"For six years, Mr. President, I have carefully watched the progress of this disease, and it really warms the recesses of my heart to find myself surrounded, as I do to-day, by the highly honorable and respectable throng of gentlemen who now grace this rostrum,—yourself, Mr. President, prominent among this galaxy of talent, education, genius, morality, and thrift!" (Immense applause, during which the speaker removed his outside coat.)

"The day is auspicious, Mr. Barnum,—I beg pardon—Mr. *President*. The spirit of liberty,—of American liberty,—sir, is abroad! To be sure, our valued friends who pretend to Know Nothing (and whose pretensions none of us here, I think, will gainsay) have commenced an onslaught upon almost everything of foreign extraction; but they kindly permit us to import Chinese *fowls*, and allow us tó breed them — for the present, at least — without interruption; for which I trust they may receive a unanimous vote of thanks from this American *National* Poultry Society." ("Yes, yes!" followed this allusion, with hearty cheers.)

"I repeat it, sir,— the times are auspicious. Money is a drug in the market, plainly. The patronage bestowed upon this show (in which, Mr. President, I am sure your native modesty and national patriotism cannot suffer you to feel the slightest *personal* interest) is evidence of this fact. The prices paid here, in 1854, for domestic fowls — though so clearly below their actual value!—supports this assertion: and your own entire lack of backwardness in coming forward to assume the risk and responsibility of the expenses of this exhibition is the crowning proof that *l'argent* is plenty — somewhere, at least. I have no disposition, Mr. President,— far be it from me — Heaven forbid that I should attempt — to offer one word of flattery, that you might, by any possibility, appropriate personally. No, sir, — I am no such man! But, if ever there was an individual whose pure-bred disinterestedness, whose incomparable generosity, whose astonishing sacrifice of self, stuck out like a sore thumb, these attributes have now been evinced, beyond the shadow of a shade of question, on this exhilarating occasion, through the astounding liberality of a gentleman, the initials of whose name are Finnyous Tee Barman!" (Immense applause, during which the Young' Un laid aside his dress-coat, and took off his cravat,—while the President, with both hands over his face, sat overpowered with his emotions.)

"Mr. President, I am no clap-trap orator. I shall say what I have to say, sir, to-day, without any hope or aim towards future reward. To be sure, I have the originals of the finest-blooded fowls in the land, and nobody disputes it; and I have now a fine lot here to dispose of; but this is not the time or place to allude to this matter; and I will only say that I do not charge so much for them as many breeders do, while, at the same time, mine are very much finer and purer than anybody else's, as can readily be seen upon examining the contents of my cages, in the first room below this hall, on the right-hand side as you enter the building. The people, sir, are in search of information on this interesting subject; and I will only add, gentlemen,—call as you pass out, and judge for yourselves." (Loud cries of "We will!—we will!" "That 's true!" "That 's a fact!" "Your fame is firmly established!")

"Mr. President, I have been too long a resident of these United States—I am too old a citizen of this enlightened country— to be ignorant of the true character of the American people. I am a Yankee, sir! My father was a Yankee, and my grandfather (if I ever had one, sir), before him. 'The people' know what they are about. You cannot deceive *them*, sir, as you and I well know. When they undertake a thing, it must go forward. There 's no stopping them, sir. They enter into any enterprise that promises so much of universal success to the whole country as does this business of poultry-raising, with a rush, sir! And they carry out their objects,— *nil disperandum hic jacit est glorii mundi morning*, sir,— as the poet remarks." (Hurra! Hurra! "Three cheers for Burnham," suggested the President, which were given with a will; and during which the speaker removed his vest and braces,— carefully securing his watch, however, at the same time.)

"We are not here to be humbugged, sir, nor do we aspire to humbug anybody, at this exhibition;— a performance which would be rather difficult to effect, in my humble judgment, even if we did! We come here to show the people what has been done, what is now doing, and what may be done again, sir, by our friends here, all of them and any of them, who choose to undertake the pleasing and delightful task of rearing *pure*-bred fowls. And, should there now be within the sound of my voice any lady or gentleman who has never seen the tiny Shanghae chick as it emerged from its delicate prison-shell and leaped forth into liberty and the glorious sunlight,— should any one of my listeners never have enjoyed the dulcet tone of that chicken's tender 'peep,'— if any of you are strangers to the habits and beauties and innocence of these rare but graceful birds,— if you have never listened to the melody of their musical crow, from youth to green old age,— I will only say, procure some of the genuine specimens, and there is much of joy and happiness yet in store for yourselves, your wives, your children, or your friends,— if you chance to have any." (Applause, and marked sensation.)

"Mr. President, I am no speech-maker. Had I, for one moment, supposed that *I* should have been thought of, by this talented and well-informed audience, I should not have been present here, I assure you. But, sir, my fame preceded me here. I'm a poor but honest man; and modesty, sir, that native modesty which so preëminently characterizes your own composition, Mr. President (had I suspected that I should have been called upon), would have prompted me to have left to others the pleasing task of speaking of me and mine. Still, if my friends '*will* buckle fortune on my back, whether I will or no,' I can only say that I feel impressed that the duty and moral obligations I owe to society compel me to submit to the burthen, with the best possible grace at my humble command." (Deep sensation among the audience; the ladies, for the most part, in tears.)

"But, sir, the future is before us! The brilliant star of fortune still shines in the distance, for the encouragement of those who have not yet availed themselves of the splendid promise that awaits the men who are yet to come after us, to do as *we* have done! And, to those who are now about to undertake the commendable occupation of attempting to breed 'fancy poultry,' I will only say, 'Go on, gentlemen! Forward, in your delightfully pleasing and profit-promising ambition! Purchase none but the best stock, without regard to price; and *breed* it (if you can!) Everybody wants to buy,—everybody *will* buy,—and the hens that lay the golden eggs are still for sale, within the sound of my voice (unless they have all been bought up since I entered this hall). But there are still a few more left, I have no doubt, gentlemen; and, I charge you, seize them while you may!'"

A general stampede followed my speech. I secured my clothes, and, for three hours afterwards, I found it impossible to get within fifty feet of my show-cages, in consequence of the throng of purchasers that crowded around them!

There must have been some charm about those magical coops of mine. They were filled and refilled, twenty times over; but they were as often emptied, and at singularly gratifying prices, both to buyer and seller.

CHAPTER XXXII.

A "CONFIDENCE" MAN.

TOWARDS the close of this show in New York, a somewhat noted cattle-breeder (who was then absent in England) wrote home to an agent in this country, directing him to secure all the Grey Shanghaes obtainable, and further to contract for the raising of hundreds or thousands more, to be delivered during the following season.

At this late day, such an undertaking appeared (to the initiated) to exhibit a most extraordinary confidence in the reality of the hen-trade; but, to those who "had been there," it was very amusing to witness the new-born zeal of this curiously verdant purchaser, who invested so large an amount of money, in 1854, in this hum!

The most extravagant prices were paid by this person for Grey fowls, and large orders were given by the agent, to different breeders, in New England, for future supplies. Several hundred birds were then purchased, at rates varying from four or five dollars to fifty dollars *each;* and finally some twenty cages were filled, and consigned to London, to be disposed of (as it was supposed) at enormous figures.

This speculation was a total failure. The fowls were inferior, and sick, and worthless. An auction sale followed quickly upon their arrival in England, the proceeds of which failed to pay even their freight and expenses out from this country; and the "confidential" proprietor of the stock, who had not the slightest conception of the details of the trade, was the loser of hundreds of dollars by this foolish and reckless undertaking.

But his contracts with home breeders, who had raised for him one hundred, three hundred, or five hundred pairs of chickens, each, were yet in *statu quo!* Two or three thousand Grey chickens were awaiting this confident gentleman's orders, and in the mean time were devouring huge quantities of corn and meal, then ranging at from a dollar to a dollar and ten cents a bushel!

Sales were merely nominal; buyers of fancy fowls were *no*whar; grain continued on the rise; the chickens grew longer in the legs and necks, and devoured more corn than ever; cold weather approached, and the breeders had no conveniences for housing these thousands of monsters; and finally the victims became importunate.

The contractor did n't want the fowls. Of course he did n't. He had "put his foot into it" with a vengeance!

But the parties who had raised these birds "to order" insisted upon the fulfilment of the contractor's promise to take them, at four, six and eight months old.

But the confident gentleman, who, in the spring of 1854, had made up his mind that the "hen fever had but just then made its appearance, in fact," *now* discovered that the bottom had been shaky for a twelvemonth, at the least, and had at length fallen out altogether!

The folly of this enterprise was apparent to every fowl-raiser in New England, from the outset. But this man knew what he was about,— so he declared,— and he scouted the advice of those who, from long experience, were able to instruct and advise him better. It was but a single instance of its kind, however, and it served, for the time being, to aid in keeping up the excitement of the humbug which had cost so many men before him large sums of money, and months of labor and care, without the slightest subsequent compensation.

By the fall of 1854, the price of this "fancy stock" began to approximate towards its intrinsic level, somewhat, in consequence of its being thus overdone; and very fair birds were offered for five to seven dollars the pair, with but few purchasers.

In England, the fever had subsided. During the spring and summer, my own sales for that market had been continuously, and without any abatement, extremely liberal;

but the prospect suddenly became clouded — the demand fell off — and I saw that the gate was about to be shut down.

The jig was nearly up, evidently, in December, 1854. In all the suburban towns of this state, and more especially throughout the entire length and breadth of Rhode Island and Connecticut, immense numbers of the Chinese varieties of fowls were being bred; and I saw, months before, that the market must of necessity be glutted, to the full, in the winter that was then approaching. Many of the experienced fanciers still clung to the hope that the trade would rally again, however,— but I was satisfied that the engine-bell had rung for the last time, and that the train was already now on the move.

A "PURE-BRED" SPECIMEN, IMPORTED FROM BRIGHTON.

CHAPTER XXXIII.

THE ESSENCE OF HUMBUG.

During this and the previous years, some of the older fanciers and breeders had resorted to the most fulsome and nonsensical style of advertisements, to push off their stock upon the unguarded. No quality of superlative goodness, known or unknown, that could be described in the English language (either by means of "communications" through the public prints, or by ordinary forms of advertising), was omitted to be proclaimed by the owners of fancy stock, in order to force off upon the credulous or the uninitiated their "newly-imported" stuff, and its progeny.

High-sounding but most ridiculous titles were given, by the nominal "importers," to their live stock; and the public were asked to purchase "Hong-Kong" fowls, "Bengal Eagle" chickens, "Wild Indian Mountain" hens, "Whang-tongs," "Quittaquongs," "Hoang-Hos," "Paduas," etc.; and the following advertisement appeared, finally, to cap the climax of this inexpressibly stupid nonsense. It was printed in an agricultural monthly, issued somewhere in western New York, and it ran as follows:

"MORMANN & HUMM, Importers and Exporters of, and Dealers in, all breeds and varieties of Blooded Live Stock, Big Falls, N. S. Messrs. Mormann and Humm are now perfecting their arrangements for *importing* from Europe and Asia all the best breeds of Horses, Cattle, Hogs, Dogs, Sheep, Rabbits, Goats, Fowls, &c. &c., and for *exporting* Buffalo, Elk, Deer, Moose, Badgers, Bears, Foxes, Swifts, Eagles, Swans, Pelicans, Cranes, Loons, &c. &c. They will keep on hand, as near as may be, all the best Blooded Animals and Fowls — gallinaceous and aquatic — fancy and substantial — which they will furnish to their numerous patrons in Europe and America at reasonable rates. All orders should be directed to Big Falls, N. S., until otherwise notified.

"Also, they have imported the finest and only PTARMAGINS ever introduced into the United States. These surprisingly beautiful fowls are direct from the original stock. The Ptarmagins — white in winter and ash-colored in summer — booted and tufted — are the most unique of domestic fowls. They will supply orders for Ptarmagin chickens; also, Hoang-Hos, Imperials, Falcon-hocked Cochins, (!) and a large variety of Improved Suffolks and other fine hogs, from the choice stocks of His Royal Highness Prince Albert, His Grace the Duke of Beaufort, Lord Wenlock, the Earl of Radnor, late Earl of Ducie, Rev. Mr. Thursby, Mr. Garbanati, &c. &c. Also some choice Chinese Mandarin and Siamese hogs, &c. &c. &c."

In this same pamphlet, appeared the annexed communication (in the form of a letter to the nominal publisher), which will explain itself, probably, to those who are acquainted with its hifalutin author. It was a rich "card," in the estimation of the "boys," at the time of its first appearance, though nobody ever saw this extraordinary beast or its progeny, I imagine :

Chinese Mandarin Hogs.

" ———, *Nov.* 7, 1854.

"FRIEND M—— :

"We have just purchased the lot of *Chinese Mandarin Swine*, imported, &c. &c. &c. * * *

"This is the best breed of China hogs, and are great favorites with the inhabitants, *the meat being remarkably tender and fine-flavored.* At maturity they weigh from fifteen to eighteen score, and are very prolific.

"The head and face of these animals very closely resemble an elephant, both as to the appearance of the skin and ears, and the number and depth of facial fissures; perfectly unique, and strikingly oriental in capital aspect.

"The neck is longer than that of anything of the hog race, imparting a most singular appearance to the proportions of the whole animal.

"These Chinese hogs are entirely different from anything of the sort ever imported into this country before, and are the most prolific of the swine race. The imported sow and each of the sow-pigs have *eighteen* well-developed dugs. The number of well-defined dugs is always the best *prima facie* evidence of prolificness in any animal.

"The bodies of these hogs are shaped like the *white* Berkshire breed of England. They take on fat with remarkable rapidity, and, in color, though not so spotted as the leopard, these hogs are beautifully striated, the body spotted like polished alabaster and ebony, checkered and rounded most exquisitely.

"We shall have an engraving of these animals for the northern agricultural papers, and one of the great English periodicals. Yours, truly, —— & ——."

The editor adds, cautiously, "The importers are gentlemen of strict probity and honor, so far as our knowledge extends; but, in these hurrying times, when the public excitement is up on any kind of stock, a man *may* import and sell worthless animals, to a great extent, before a reäction can take place."

Now, this sort of mush and moonshine very soon nauseated upon the stomachs of "the people," even; who ordinarily can (and will) patiently submit to a vast deal of mummery. But when such palpable bosh as this is placed before them, they are apt to dodge all association with it and its clearly-expressed humbuggery; and so the tide now very quickly began to turn against the trade. "Brahmas," and "Quittaquong" fowls, and "Mandarin" pigs, proved too threatening a dose for the masses! They had n't time to spell out the names of such stock — to say nothing of purchasing it, at round figures, and attempting to *breed* it afterwards.

What those men imagined they could possibly effect by this sort of ridiculous nonsense, I am unable to conceive of. Yet it was put forth in sober earnest; and scores of similar advertisements filled the papers, from time to time — each having for its object the continuous gulling of the "dear people," each in its own peculiar way.

And for years — up to this period — the star-gazing, wonder-loving, humbug-seeking portion of the community,

— the mass who fill every corner of the land, and who watch for something continuously "new under the sun," out of which money can be made,— I say, for years, this portion of the public believed what they saw and read of, and responded to this sort of thing with a gusto equalled only by the zest with which, in years before, they had encouraged and supported the score of other "hums" that had been current around them.

But the delusions of morus multicaulis, and Merino sheep, and patent bee-keeping, and Berkshire pigs, and tulip-growing, had passed away; and the hen fever, at last, subsided, too. Unpronounceable names and long-winded advertisements would n't do! "The people" had ascertained that there was an end even to Shanghae and Brahma-ism! And this flimsiest of *all* bubbles was now inflated fully to bursting.

CHAPTER XXXIV.

A TRUMP CARD.

Not to be beaten by this sort of thing (since the columns of certain friendly journals were still open to me), I adopted the style of advertising then current; and soon after the articles noted in the last chapter made their appearance in the "agricultural" paper alluded to, the following letter from the Young 'Un was published in the New York *Spirit of the Times*, upon the subject of live stock generally, and what *I* had for sale particularly.

"Uncle Porter:

"During the last few years, I have turned my attention to trafficking in stock (as you may *possibly* already be aware). Not copper stock, or Reading, or Hoosac Tunnel, or similar 'bores,'—but in *live* stock; to wit, living stock. As is usual in this great and free country, other people have got to doing the same kind of business, since it has been now found to 'pay;' and who 's a better right?

"*I* desire, at the commencement of the new year, through the *Spirit*, to call the attention of such of *your* friends (as you cannot supply readily) to my present assortment of *ominus*, omnivorous, carnivorous, graminivorous and bipederous specimens — which I have imported from Europe, Asia, Africa, Oceanica, South America, and *other* places; and consisting, *in part*, of the following, namely:

"All the best and choicest breeds and varieties of horses, cattle, swine, dogs, cats, sheep, rabbits, goats, fowls, pigeons, rats, catamounts, hyenas, alligators, cormorants, kangaroos, grizzly bears, antelopes, envelopes, llamas, lam-'ems, jaguars, fox and geese, kinkajous, petrel, periwinkles, long-tailed rabbits, Nubian fennecs, red eagles, condors, hooded ducks and hood-winked drakes, swifts, sloes (intended for 'fast' men and old 'fogies'), chamois, armadilloes, wingless emus, beadles, crabs, cranes, coons (bred from 'that same old 'un'), white zebras, macaws, catspaws, cantelopes, carbuncles and shuttle-sewing machines.

"I also have, for *exporting*, a splendid assortment of buffalo, elk, deer, moose, bears, cranes, owls, badgers, woodchucks, swans, pelicans, gulls (genuine), rattle-snakes (domesticated), fighting hen-turkeys (from Iowa), larks (from Nauvoo), and a superior assortment of *fishes*, of every conceivable size, color and variety, which are warranted to live out of the water, in any climate. In short, I will keep on hand all the best 'blooded' animals, fowls, quadrupeds, fishes, reptiles, insects and birds,— be they gallinaceous, aquatic, aërial, fancy, substantial, good, bad or indifferent, that may be had; which I will furnish to my numerous friends, patrons, and the rest of mankind, in Europe, Asia, Africa or America, at all hours of the day or night (Sundays excepted); and at prices so reasonable that Christendom shall 'vote me' a philanthropist, or no sale.

"Among my most recently received samples, I beg especially to call the attention of fanciers, amateurs and breeders, to a 'vaggin-load of monkeys, vith their tails burned off,' which I warrant will not frighten the most skittish of horses. A crate of she-basilisks, of most virtuous exteriors, and with eyes as large as saucers. Eleven pet elephants (intended to have been offered to Mr. Barnum, but who informs me that he has done breeding them, on account of the high price of provender). One pair of red ostriches,—supposed to be the original progenitors of the famous 'Cochin-China' race of poultry. (The male has a 'horse-shoe mark' upon his breast, described by certain modern authors on poultry. Unluckily for this theory, however, I happen to know that this individual was kicked by a mare of mine, while the beauty was skulking behind her, and attempting to rob her of the corn she was eating from her crib.) I have a trio of very healthy walruses, from Norway, that will eat snowballs from your hand. Also, a brace of young mastodons, very docile, and as easily kept, *almost*, as a trio of 'Brahma Pootras.' Three *green* swans (delightfully green), that never seek for or approach the water; supposed not yet to have learned to swim. I have also in my collection a family of very curious chameleons (believed to be), but none of which are supplied with the usual caudal extremity yclept *a tail.*

"My friend Durivage — who, as you are aware, is now in the Boston Custom-house, and whose opinion, consequently, is n't worth much — examined this family, and at once pronounced them hop-toads! But I don't mind *his* jokes. *You* must see them. They are beautiful creatures, and '*do* live on air,' I assure you; I have seen them do it frequently, without changing color. Dr. Bennett, of Fort des Moines, has recently sent me a fine male porcupine,— a nice little fellow to handle, so long as you rub his feathers the right way,— which I purpose to cross upon my Chinese Mandarin sow, at a future day, for experiment. In addition to all these, I have, of *fowls*, the Mum-chums, Hong-Kongs, Whamphoas, Quittaquongs, Hoanghos, Bramapooters, Damphules, Rocky-mountain-Indian-wharhoops, Nincompoops, etc., and an endless variety of white blackbirds, sleeping weasels, very fine mules (for breeding), fan-tail tumblers and tumbling fantails, no-woolled sheep, etc. etc., and so forth.

"The principal object of this communication, however, is *not* to particularize my stock, but rather to call attention to my new breed of Hogs, which I have lately imported; and of which I send you a striking likeness herewith. I call it the Chinese Mandarin Hog.

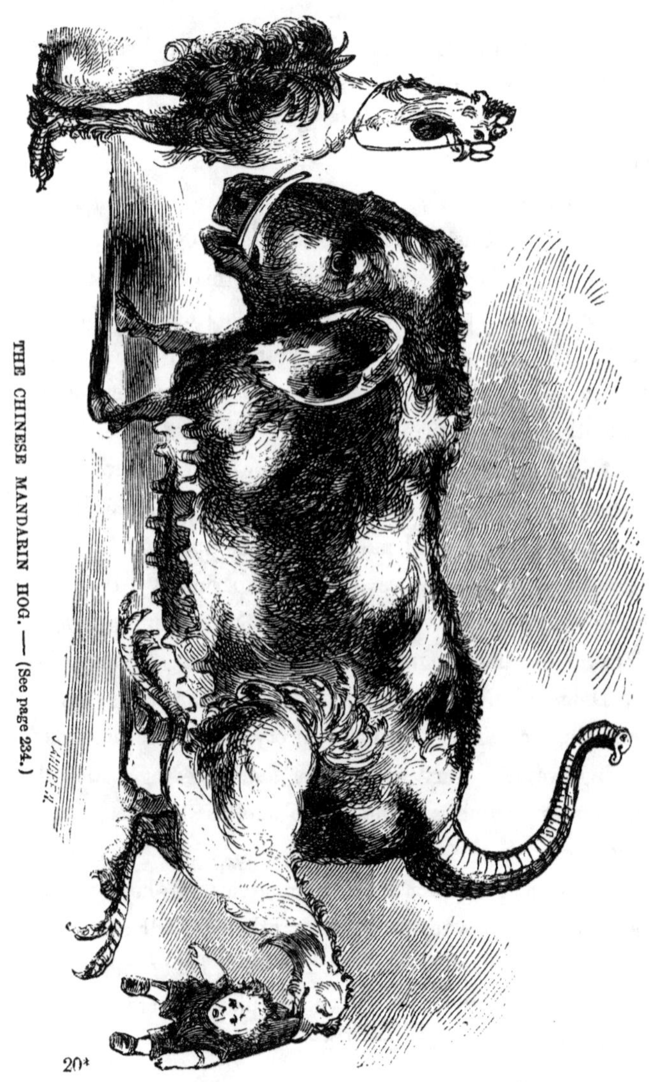

THE CHINESE MANDARIN HOG. — (See page 234.)

"The drawing of this very faithful and life-like picture — copies of which I have already forwarded to *Punch*, the Paris *Charivari*, etc.— was executed by Phizz; the engraving is by Quizz; the portraits are perfect.

"This breed of hogs is most extraordinary; and has been pronounced of great value for their beautiful model (see portrait), and easy fattening qualities. Their meat is also remarkably tender and fine-flavored, as can be proved by several gentlemen in this country, although this is the *first* hog of the kind ever brought here, and she is now alive! As you will note in the drawing, the head and face of these hogs (supposing it possible that another could be found on God's footstool of the same kind) very closely resemble an elephant; perfectly unique, and strikingly oriental in capital aspect. (Which, if you do not understand, I can only say is plain English, and I must again refer you to the picture.) There is another singular feature, you will probably have observed (allowing that you are somewhat acquainted with the *ordinary* formation of animals), and that is, that the trunk of this animal is upon the wrong extremity; but it answers, apparently, a very good purpose for a tail, as will be noted. True, the neck is longer than that of any hogs ever seen here, imparting a singular appearance; but it is a long lane that has no turn in it, and so *n'importe* on this point.

"This is the most *prolific* of the whole swine race.

There never was one in America before, but this point is settled. She has eighteen dugs (see portrait), and learned doctors inform us that the number of dugs (teats) is always evidence of prolificness. The bodies of these hogs are like the *white* 'Berkshires' of England (admitting that the white and the black Berkshires have different-shaped *bodies*). In color, though not so spotted as the leopard, these hogs are beautifully striated, like polished alabaster and ebony, checkered and rounded (see drawing) most exquisitely, like a slice of mouldy sage cheese.

"P. S. Although I am now short — or shall be, in the spring — full eleven thousand pairs of pigs, from this sow (to fill present orders), yet I will undertake to furnish a few more to gentlemen who may fancy them, at the advanced price,— seven-and-sixpence per pair. (I have no *boar* of this breed, but that is immaterial.)

"N. B. I have frequently been asked to account for the singular facial appearance of this sow; but I can only do so, satisfactorily to myself, upon the theory of my friend Jacob, of old; that, *at some time or other*, her mother must have 'seen the elephant'!

"*** The other figures in the accompanying drawing are likenesses, also from life, of my harmless and beautiful 'Bramerpootrers.' They are very fond of little children (see picture), and I send to my uncle William Porter, herewith, as a New Year's Gift to our mutual friend, Solon

Robinson, a very fine sample, with the gentle hint that if he keeps his 'Hot Corn' as far out of this fellow's reach as it has thus far been out of mine, it will be perfectly safe.

"☞ All orders for my famous 'Bramerpootrers,' or my imported 'Chinese Mandarin Hogs,' etc., must be put in water-proof condition, post-paid, endorsed by the collector of this port, and sent, by Adams & Co.'s Express, to Niagara Falls, until I conclude to remove to Salt Lake, Nebraska, or 'elsewhere,' of which due notice will be given (provided I don't decide to 'step out' between two days). *Adios!* Yours,

"*The* YOUNG 'UN.

"*Boston, Jan.*, 1854."

Now, the above letter explains itself fairly, upon its face; yet — would it be believed? — I actually received four or five sober (I *presume* the writers were sober) letters of inquiry, relating to the "curious and remarkable Chinese Mandarin Hog in my possession," immediately after the above article appeared in the *Spirit!* Such are the knowledge and acquirements of "the people," in certain quarters, upon the subject of live stock!

CHAPTER XXXV.

"HOLD YOUR HORSES."

My competitors in the hen-trade, by this time, had got to be exceedingly active and zealous, though they rarely indulged in personalities towards me, at all. Generous, disinterested, liberal, kind-hearted, valiant men! Providence will reward you all, I have no doubt, *some day or other!*

The following article, which appeared in a "respectable" agricultural sheet (which, though I was solicited so to do, I neither subscribed for nor advertised in), I offer here as a sample of the puffs that were extended to me for five years, by the small-fry humbugs whom I rarely condescended to notice. This "elegant extract" appeared in a northern *Farmer:*

"We did suppose that the strait-jacket we fitted to this fellow (Burnham) would be worn by him, but it appears that, on reading our article relative to his movements in England in regard to Grey Shanghae fowls, he cast it off, and made an attempt to put us *hors du combat*, in his usual style.

"But we must say that his pretensions to being an '*importer*' of these fowls, to having the '*original*' stock, to being the importer of the fowls he sent to England, is the greatest deception that ever came under our observation. But this is only in character with the general transactions of the man. In his dealings generally he seems to have had no other object in view but to *get all he could for his fowls*, with no regard to their *merits*. This is shown by a letter of his, which we have in our possession, written in 1852 to Dr. Bennett, in which he uses the following language, in regard to fowls: 'Anything that will *sell*,—bah!'

"We will take the liberty to digress a moment, to make a few remarks on his penchant for the use of the expression '*bah!*' which is his common habit in correspondence. When Burnham was a loafer at large, previous to his *foul* speculations, it is said that he was very fond of *mutton;* and as many a fat lamb was missed in the vicinity where he resided, it was more than suspected that he knew what became of them. Whether this be so or not, it seems that '*bah*' is ever escaping from his lips, a judgment, as it were, for the alleged iniquity of disturbing the nocturnal peace of that quiet animal. * * * *

"Now, friend Burnham, do be civil and *honest*. Your having sold 'premium' Cochins all over the country, with the *real* 'premium' fowls in your own yard, will soon be forgotten, and you may yet be considered a clever, honest fellow; but you *must* stop pretending to be an 'importer' of fowls."

I was thus charged with putting my "friend" *hors du combat*, with lying generally, with sheep-stealing, with selling "premium" fowls over and over again, as well as with striving *to get all I could for my poultry*,—this *last* offence being the most heinous of all! But, as I lived (as I supposed I should) to see this cub and his allies on their knees to me (as I could show, if I desired to do so, now), I did not mind these first-rate notices. They were most decidedly of *miner* consideration in my esteem, when I thought how "the people" crowded around me to obtain eggs or samples of my famed "imported," "superior," "magnificent" and "never-to-be-too-much-lauded" pure-bred fowls!

In the official Report upon the first New York show, the Committee of Judges there state that, "though they have been governed by the nomenclature of the list, they by no means assent to it as a proper classification. *Shanghae* and *Cochin-China* are convertible terms, and Brahma Pootra is a name for a sub-variety of Shanghaes, of great size and beauty. White *Calcuttas* and *Hong-Kongs* were not on exhibition. Believing them to be inferior specimens of White and Black Shanghaes, it is likely that we would not have awarded them premiums, if found. In lieu thereof, we have assigned several additional second premiums for *Brahma Shanghaes.*

"For the sake of simplicity, we would recommend that *all* thorough-bred large Asiatic fowls be classed under the name of *Shanghae*, to be further designated by their color; and, inasmuch as these shows are intended not solely for the aggrandizement of breeders, but for the purpose of converting 'Henology' into a science, we would earnestly suggest that all ridiculous, unmeaning *aliases* be abandoned, and a simple, intelligible and truthful classification strictly observed." After quoting this, the writer above alluded to objects to the recommendation to call *all* Asiatic fowls *Shanghaes*, notwithstanding the action of the Committees of the National Society. He insists:

> "This is a ridiculous affair, and we call on fowl-breeders to *veto* this nonsense at the outset. Just imagine what a ridiculous figure breeders would cut in calling their fowls '*Brahma* Shanghaes,' '*Chittagong* Shanghaes,' &c.! Why this desire to overturn *established* names? It arises from a *prejudice* against the *name* 'Brahma Pootra,' and a desire to *put down* that popular breed. Again: *Who* are the gentlemen who recommend such a course? Why don't they give their *names?* These 'recommendations' and 'resolutions' are no more the act of the *National Poultry Society* than of the Emperor of Russia! Where were the *forty* MANAGERS when the above 'resolution' was passed? *We*, as *one*, were not there; and we learn that not over *three* out of the entire number were present, and that the resolution was passed by *outsiders*, and, perhaps, influenced to do so by G. P. Burnham, of '*Grey Shanghae*' notoriety."

This clown even "regrets that he did not attend this show;" as if it would have made a difference in the result! Well, well! — the impudence and ignorance of some people really astound us, at times! He says "some of the best Brahma Pootra fowls were entered 'Chittagongs.' Now, we declare emphatically that the desire on the part of certain breeders to class the *Brahmas* as identical with the *Chittagong* fowl is absurd; and we assert that no man can produce any evidence that the Brahmas are identical with Chittagongs, beyond the fact that many breeders have produced mongrels, by crossing Brahmas with Chittagongs, and now seek to amalgamate the two breeds."

Who ever wished to "produce any evidence" on this subject, pray? "The people" wanted *fowls;* they never sought for "evidence," man! The breeder who could "produce" fowls was the man to succeed in the hen-trade. As you never did this, and only bought and sold wretched mongrels, with long names, you never succeeded. And "the people" said, "Served you right!"

This sapient editor then declares that he "doubts the ability of any *Poultry* Society to maintain its existence *permanently*, for the reason that such societies will, sooner or later, degenerate into mere *speculating* cliques, and the premiums will become a matter of *barter*, or a matter of *favor* to particular men, like the operations of our government."

Is it possible! When did you discover this extraordinary and singular fact, my dear sir? Not until the close of the year 1854! After the cars had long since passed by, and the fun was over, effectually and forever, in this country. Your warning was valuable, indeed! The colt had left the stable, and you *now* come to fasten the door! O, chief of prophets in Henology! how much "the people" owe you for your advice and foresight in this hum!

This writer finally thus wriggles over the action of the "National" Society at New York, which knocked his "Bother'ems" on the head so effectually, substituting their true name (the "Grey Shanghaes") for this ridiculously assumed cognomen. He continues:

> "The most absurd thing which came under our observation at the fair was the *classification* of certain fowls. There were the beautiful white Brahmas, with pencilled neck hackles, placed by the side of fowls of an owl or hawk color, and both classed '*Grey Shanghaes!*' How long will a few old fogies thus stultify themselves? Many exhibitors were highly displeased with this absurdity. They who think that the name of Brahma fowls can be changed to 'Grey Shanghaes' have entirely mistaken their ability to make such an innovation. What did all the nonsense in the resolutions passed at the National Poultry Show in New York about the nomenclature of fowls effect? Just nothing at all."

Indeed! Did n't it? Is it possible? You don't say so! My dear friend, you have a great deal to learn yet; and I here advise you, affectionately and lovingly, and with an ardent desire for your present and future good, to — "hold your horses!"

CHAPTER XXXVI.

TRICKS OF THE TRADE.

POULTRY exhibitions had been or were now being held all over the country. In the New England States, in New York, Ohio, Pennsylvania, Maryland and Virginia, numerous fairs had come off, at which the customary competition among breeders of fancy poultry had been duly shown; and for a time, yet, out of Massachusetts, the fever still raged, though with comparative abatement.

It was now a common thing, and certain men were in the habit of visiting the express offices, and examining coops of fowls, and taking the names of the persons to whom they were directed, and then writing them that they would furnish such fowls at a much cheaper rate. This occurred, generally, while the stock was *en route* to its destination; but it never disturbed *me*.

Among the Rhode Islanders (who, by the way, generally speaking, have raised the best of all the Chinese varieties of fowls, for five years past) a feeling of desperate rivalry had grown up. At the Providence shows, many of the choicest specimens ever seen among us were exhibited and disposed of at high rates. But the management of the fairs there was not satisfactory to certain breeders, who, unfortunately, and naturally, drove rather "too slow coaches" to keep pace with a few of the leaders in the traffic there, as will be seen by the following *exposé*, which I find in the shape of an advertisement in the *Woonsocket Patriot:*

In a report published subsequently to this State Fair in Rhode Island, the Committee on Poultry at the exhibition held there in the fall of 1854 awarded their first premium to the *chairman* of the committee. The second premium was awarded to another man, who had just as good fowls, probably, but who was n't smart enough to "keep up" with his competitor. The person who came out thus second-best, only, at once charged, through the public prints, that an attempt had been made by the chairman thus "to hoodwink the public" in their future purchases (which was very likely, because it was a very common matter). The injured party says, in his published "card,"—

"No doubt Mr. C—— was ready to grasp at the appointment as the committee, and he was progressing in the examination, when I remonstrated, and had two other men added to the committee with him, supposing that justice would then be administered to the parties concerned. But Mr. C—— was determined to have the sole arrangement of the report, contending with the other two upwards of five hours, aggrandizing to himself the first premium, and then affixing to the committee's report the name of Mr. A——, instead of his own, to deceive the public, that he was not interested. Mr. C—— intended that justice should not be done his competitor, by withholding his right as to the first premium; and I challenge him to an impartial exhibition of the poultry (although some of his number were borrowed), for the sum of one hundred dollars, to be decided by three disinterested men."

Another member of this committee then states that, "being one of the Committee on Poultry at the late State Fair, held in Providence, R. I., and having seen the report of the same, I feel it my duty to say that such was *not* the decision of the committee. Two were in favor of giving to —— the *first* premium; as we could not agree, we decided to award a premium of *twelve dollars* to ——, also the same to Mr. C—— provided each were represented equal in the report."

Now, this was a very trifling affair to trouble the public with, yet it shows "how the thing was done." Mr. C—— had a happy way of "laying 'em all out," when *I* was not in the field. If the advertisements "to the public" were paid for duly (and I presume they were), I have no doubt the public are satisfied; and Mr. ——, the injured party, must keep his eyes open tight, if he trains in company with experienced hen-men. This is but "a part of the system," man!

Now, as this sort of thing was of very common occurrence among the hucksters who kept the hen-trade alive, for years, this was in nowise a matter of astonishment to the "hard heads" in the business. The only wonder was that the man who performed *this* trifling trick did not carry out the dodge more effectually, and bear away *all* the premiums in a similar manner, as had been done by some of his smarter predecessors!

The editor of a New York journal undertook as follows to "inform the public" (in 1854) of a little performance in kind, which had been common for several years at these fairs where "premiums" were awarded, and which proved a very profitable mode of operation, almost from the very beginning of fowl-shows in the United States. In an article upon a recent exhibition, under the caption "*How the Cards are Played*," he says:

"A fowl-breeder, by extraordinary means, raises a *few* specimens of fowls of great size, which he takes to the exhibition; and, on the appearance and character of those *few* specimens, he contracts to furnish fowls and eggs of the 'same stock.' He goes home with his pockets full of orders, and with not a *single fowl, for sale*, in his possession at the time, and hastens to purchase of A, B and C, such fowls as he can find, say at $3, $5 to $10 a pair, which he sends to fill his orders at $20 to $50 a pair, and no nearer in value to the stock that appeared on exhibition than a turkey is to a turkey *buzzard!* The same of *eggs.* Now, there are exceptions to this allegation. but we KNOW that such things are done, and we think that the public should be put on their guard."

There is no question about the accuracy of this statement. The writer says he "*knows* that such things were done;" and I feel sure that no man in New York State ever knew the details of this dodge so well as *he* did. It was a very common thing everywhere, however, among the hucksters. I had no occasion to resort to this plan; for the game *we* played was a deeper one, altogether.

There was a "live Yankee," all the way from Rhode Island, who attended the New York show, who took the boys down there after the following style. as appears from another advertisement, which I recently met with, and which feat is thus described by one of the sufferers. In a "card" published soon after that exhibition, this victim of misplaced confidence says, with a show of seeming injured innocence:

"Justice to the public, as well as myself, demands a slight explanation of a few facts connected with the recent National Poultry Show, in New York City.

"Mr. C——, of Woonsocket, R. I., accompanied me to New York for the purpose of attending the fair. On the fourth day of the exhibition it was announced that the judges were about to commence their labors. Mr. C——, seeing that his chance for a premium of *any* kind on Asiatic fowls was very slim, came to me and requested, nay, even *insisted*, on grounds of mutual friendship, that I should put my two best hens with a cock of his, for the purpose of taking the first premium. I finally consented, with the ex press understanding, *and no other*, that we should each share the honors and proceeds equally. On Friday it was announced, in the lecture-room, that *he* had taken the first premium on the best pair of Asiatic fowls, of whatever sub-variety. I went to him, at once, and expressed my dissatisfaction, and reminded him of his agreement. He then agreed to see the secretary and all the reporters, and publish, or cause to be published, a card, stating that I was equally entitled to the premium with himself, as the hens were raised by me; and he furthermore agreed that his name should not be mentioned or published, in relation to the premium, except in connection with my own. How was that agreement fulfilled? On taking up one of the New York dailies the next morning, I was surprised to see a puff laudatory of Mr. C——, while *my* name was not alluded to,—which puff, report says, was paid for with a rooster. On my return home, a few days afterwards, I found that he had volunteered to make the following assertions: 'Well, I have laid 'em all out. I took the first premium on everything, best pair and all, and I can beat the world.' When asked how it was done, he said, 'I will tell you, *some time*, how I played my card.' "

But Mr. C——, with that reserve and indifference peculiar to gentlemen in the hen-trade who have accomplished a "neat operation," did not see fit to explain the process, and hesitated to inform his "friend" how he played his card. And so the aggrieved party resorted to the newspaper, and come the "power of the press" upon Mr. C——, as follows:

"Mr. C—— stated that my stock was 'mongrel,' and inferior. Whether it be so or not, is for the thousands and tens of thousands who saw them, while on exhibition, to judge. After selecting two of my best hens for Mr. C——'s especial benefit (as it appears), the committee *even then* saw fit to award me a premium, while his two coops of '*pure, full-blooded* Asiatic fowls,' which he had cracked up so loud and extensively, did not receive, as I can learn, even a passing notice, *except the old cock*, which was put in the coop with my 'mongrel hens,' as he is pleased to call them. Perhaps the public would also be gratified to learn the manner in which he obtained the first premium at the recent Agricultural Fair in Providence, R. I. Was it not done by entering several coops of fowls, belonging to another person, in his own name, without that person's knowledge and consent, and pointing out those fowls to one or more of the judges, representing them as his own? No doubt the books of the society, and those of the railroad corporation which conveyed Mr. C——'s poultry to and from the fair, if compared, will throw some light upon the subject. Is not this the manner in which he has frequently played his card; or, in other words, 'laid 'em all out'? As I have always treated him as a gentleman, a neighbor and friend, to what cause can I impute this low, mean contemptible and underhand manner of exalting himself at my expense? I would advise him, in conclusion, to peruse Æsop's moral and instructive fable of the ambitious Jackdaw, and learn from that, that however well a course of deception and duplicity may at first prosper, the day of exposure and disgrace will come, and the ungainly Jackdaw, stripped of his ill-gotten plumage, will stand forth in all his native blackness and deformity."

Now, I have no doubt, that this Mr. C——, when he read the above "card" (which must have cost its author considerable time and money), felt very badly about it, the more especially as the show-prizes had been duly announced, and he had the premium-money safely in his own pocket! And it certainly must have been a very gratifying circumstance, to the man who had been thus duped, to see his advertisement thus in print, too. Had *I* been similarly situated, however, after losing my premium and the credit that belonged to my having had the best fowls on exhibition, also (only by thus joining issue with another to gull the "dear people"), I rather think I should not have *published* the facts, to show myself up a fool as well as a knave. But this is merely a matter of taste. Mr. B——, who signs this "card," will scarcely be caught in this way again. We "live to learn."

Mr. B—— had not become apprised of the fact that, from the very commencement, the hen-trade was a huge gull, possessing an unconscionable maw, and most inconceivable powers of digestion. Older heads and wiser men than he had been duped or swallowed by this monster, that stalked about the earth for six long years, seeking whom he might devour. If this is the worst treatment he ever experienced at the hands of those who helped to feed the vampire, Mr. B—— is, indeed, a fortunate man. There *be* those who would gladly exchange places with this gentleman, and give him large odds.

C—— was *smart*. I have known him for several years. He is one of the few "hen-men" whom I would trust alone with my purse. And whether he raised them, or purchased them, it matters nothing; he has *sold* some of the best fowls in America.

In all human probability, the author of the "card" last quoted will live long enough (unless he shall have already stepped out) to know that "the people" went into the hen-trade blindfolded, and that the bandages have now dropped from their eyes. He will have ascertained, too, I think, that a resort to the newspapers for redress against such of his "friends" as may get ahead of his time in this way is precious poor consolation, when he reflects that advertisements cost money, and that the anathemas of an over-reached chicken-man have never yet been known to harm anybody — as far as heard from! *Selah!*

CHAPTER XXXVII.

FINAL DEATH-THROES.

THE officers and the judges at the poultry-fairs (most of whom are self-constituted), as will be seen, usually carried away all the first prizes. At a late show of the New York *State* Society, the president thereof received about *one third* of all the premiums awarded, and yet his fowls were nearly all *second* and *third* rate, and not one of them, it was stated, was *bred* by him. He may have bred a few specimens during last season, but not *one* on exhibition was bred by him. The people and certain greenhorns were astonished to see the way in which the premiums were awarded to him. One of the judges there seemed determined to award to him every premium that his influence could secure, right or wrong; and, from what was learned from exhibitors, it did look very much like an existing understanding between the parties in regard to the premiums.

For the above statement we have the authority of a huckster in New York, who did *not* obtain any premiums, and who says of the management of the state show there, that this sort of partiality shown in favor of the wire-pullers "is the rock on which the 'New England Poultry Society' foundered; and our state society is treading in the footsteps of its 'illustrious predecessor.'"

This writer contends that the president of the New York society, who thus received about all the premiums at one of their late shows, was a man of too much discernment not to see that such a *farce* as some of the judges played would redound to his discredit. They went *too far*—overdid the matter; hence the universal indignation of exhibitors. And then concludes that "poultry-societies generally merge into mere *speculating* gatherings, a *few* receiving most of the premiums, while the uninitiated exhibitor is made a tool to swell the income of those who pull the wires. Many breeders exhibit solely for the sake of the *notoriety* that their fowls will receive,—a sort of *gratuitous* advertising,"—and it is now got to be "notorious that an order sent to one who receives the *first* premium for fowls is no more likely, in many cases, to be filled with any better fowls than if sent to one who took no premium at all; as the *prize* fowls are not often for sale, and very inferior specimens are sent when orders are received."

This information would have answered very well, had it been afforded years ago. Now that the fever has disappeared almost entirely, and now that everybody has been gulled, and gouged, and *gorged*, with the fulsome and glowing accounts of the asserted reality of this thing, from the pen of this very man among the rest, it comes rather late in the day for such an one to "warn the people," and in such a manner!

But, soon after the exhibition above referred to had closed, the president of the society issued a most astounding "card," *declining* to receive the premiums awarded him, and in which appears the following sentence:

"In connection with the report of the Judges of the late State Poultry Show, allow me to make a statement. As appears from the report, my birds have been unusually successful in the contest for premiums, sixteen out of twenty distinct varieties exhibited being so honored. This was more than I expected, and more than I honestly think they deserved. And I am strongly of opinion that, had they had more time, they would have come to a different conclusion, in two or three cases."

I was prepared for almost anything in the hen-trade, up to this time; but this performance really astonished *me!* The man actually refused to take the premiums awarded him! He even went so far as to show the "judges" who *ought* to have had the prizes, rather than himself. And he actually sent back to the committee the money they forwarded to him after the exhibition was over!!

Now, if this were not sufficient to astonish "the

people," I am very much in error regarding the ordinary strength of their nerves. It was an almost immaculate performance; and the "New York State Poultry Society" should positively insist that this extraordinary man (if he can be proved to be sane) should at once accept from them one of the largest-sized leather medals, to be worn next to his gizzard, for this unexampled disinterestedness, and extraordinary sacrifice of self. O, but *that* gentleman must be "a brick," indeed!

A journal that alluded to this singular circumstance, at the time, asserted that this procedure on the part of the president "was highly commendable in the author, if his statements were made through *principle*, rather than through fear to encounter *public opinion*. He stands high in the estimation of the public, and we have ever considered him as strictly honorable in all his business transactions; but we cannot help thinking that 'a screw was loose' somewhere in the matter. His statements are not very flattering to the judgment of the judges, and show that some of them, at least, were not competent to discharge their duties properly," etc.; while, in *my* opinion, than this, a more barefaced piece of *mush* was never yet perpetrated, in the details even of the hen-trade.

This was emphatically among the "death-throes" of the *mania*. And cards like the following found their way into the newspapers, about this time, in further proof that the valve of this huge balloon had slipped out. An ambitious Western man says:

"I have long been expecting to hear of the swindling operations of a certain dealer, who makes a great display of *pretending* to have *every* breed known or bred in this country; and, to my *certain* knowledge, buys all, or nearly all, of his fowls, as wanted, and as many on *credit* as he can, but does not *pay*, nor can the *law reach* him to make him pay. I believe, also, that the papers that advertise for him are doing it for *nothing* — that is, that they are not, and never will be paid for it.

"Such a course, in my opinion, is no better than highway robbery; and I hereby give said person fair warning to act honestly hereafter, or I will point him out in a way that shall not be misunderstood, as I cannot see such rascality perpetrated, and remain silent.

"A man who deals in high-priced fowls, in receiving pay in *advance*, has his customers completely at his mercy, especially when he is not *responsible* for a copper; and at the rates that fowls sell for — say, from ten dollars to one hundred dollars a pair — purchasers should receive what is promised them,— good specimens of the *pure* breeds. So far as *weight* is concerned, a pair of fowls will fall off *a few pounds* in a journey of a week or less, in a cramped condition, and perhaps without food for a portion of the time; but in other respects justice should be done to the confiding purchaser."

Beautiful! — poetical! — musical! This advertiser, I have no doubt, keeps only *pure* stock. I do not know who he is; but, if I wanted to buy (which I don't), I should certainly apply to such an honest and justice-loving person, because I should feel assured, after reading such an advertisement, that *that* man was a professor of religion; and, even if he had the chance, would never fleece me — *over the left!*

Other fanciers, in their utter desperation (as the fever so positively and now rapidly begun to decline), resorted to the printing of the *pedigrees* of their stock; and the following advertisements made their appearance late in 1854:

" By the influence of Mr. Ellibeth Watch (editor of the *London Polkem Chronicler*, and uncle to the Turkish Bashaw with three long tails), I have just procured a few of Prince Albert's famous breed of 'Windsor fowls.' In a letter to me of the 32d day of April, Mr. Watch observes:

> 'I have positively ordered a trio of *Windsor* Fowls of Prince Albert, for you. It is THE BEST BREED IN ENGLAND, and they are much run after, and cannot be had without giving previous notice; but you are safe to have yours. I have engaged a friend to choose yours for you; and I consider it *a great thing* to get them direct from the Prince, for you must be aware that persons generally cannot exactly *pick and choose* FROM THE PRINCE'S OWN STOCK. I shall employ an efficient person to have them shipped, etc.' "

In due time this remarkable stock arrived in America, and their pedigrees were duly published; the advertiser being "thus particular," because (as he asserted) "there had been so much imposition upon the public by irresponsible persons *claiming to have made importations*" !

Now I never entertained the slightest objections to this sort of advertisement,—not *I*, i'faith! On the contrary, I deem all this kind of thing very excellent, in its way, to be sure. The more the merrier. "The people" want it, and let them have it, say I.

But, at the same time, though the "Porte-Monnaie I owe 'ems" declare that their unrivalled stock comes from Prince Albert's yards, I feel very well assured that all this is a mere guy, it being very well known that His Royal Highness is not engaged in the hen-trade particularly, and of course has something else to do besides supplying even the "Porte-Monnaie Company" with his pigs and chickens.

It was a rare undertaking, this importing live stock (with any expectation of selling it) in the fall of 1854! But we shall soon see who were the final victims of the "fever."

CHAPTER XXXVIII.

THE PORTE-MONNAIE I OWE 'EM COMPANY.

It has been said, with much of truth, that "two of a calling rarely agree;" and this applies with force to those engaged in the "hen-trade." Messrs. Mormann and Humm, whom I have before spoken of, could n't long agree together, and their "dissolution" soon appeared; and, from the ashes of the professional part of this firm, there suddenly arose an entirely new dodge, under the big-sounding title of

"THE PORTE-MONNAIE I OWE 'EM COMPANY."

The presiding genius of this concern was one Doctor Bangit,— an old friend of mine, who had been through wars enough to have killed a regiment of ghouls, who was among the earliest advocates and supporters of the "New England Mutual Admiration Society," who was one of the very first physicians employed in prescribing for the hen fever in this country, and who, I *supposed*, had had sufficient experience not to embark (at this late day) in such a ridiculous enterprise as this so clearly seemed to be.

But the doctor saw his victims in prospective, probably; and, though he had run the hummery of the fowl-fever so far into the ground that, in his case, it would surely never know a day of resurrection, still he was ambitious and hopeful; and he flattered himself (and some others) that the *last man* who bought live stock had *not* yet turned up! And so the doctor pushed on, once more.

The "BLOOD STOCK" of the "*Porte-Monnaie I owe 'em Company*"* was thus advertised, also:

"IN ADDITION to the genuine, unadulterated Prince Albert fowls, the 'Porte-Monnaie I owe 'ems' offer pigs, with tails on, of the Winsor, Unproved Essex, Proved Suffolks, Yorkshire, Wild Indian, Bramerpouter, Siam, Hong-Kongo, Emperor Napoleons, and Shanghae Breeds; most of them of new styles, and warranted to hold their colors in any climate.

"Also, Welsh Rarebits — bred from their Merino buck 'Champum,' of England (that *did n't* take the first prize at the National Show, because Mr. Burnham's 'Knockum' did!), whose ears are each thirty-three feet longer than those of our best pure-bred jackasses, and wider than five snow-shovels, by actual measurement.

"Also, A-quack-it fowls; as Swans (*Porte-Monnaie*

* I trust that this association may not be confounded with the "*Fort Des Moines Iowa Company.*" The difference will plainly be seen, of course.

I owe 'em strain), Two-lice, Hong-gong, Brumagem and other Geese. Ruin and Ailsburied Ducks, and Pharmigan Pigeons (blue-billed).

"Also, every breed of Gallinaceous fowls,— Games and other bloods already noted,— together with every species of pure and select blood-stock, which has been secured in Europe, Asia, Africa, and the Arctic Ocean, with reference to QUALITY, without regard to *price*.

"☞ We can furnish pedigrees to all buyers who desire them, which will be endorsed by the faculty of Riply College, Iowa.

"N. B. The 'Winsor' breed of pigs imported by us is a great addition to the already fine hog stock of the United States, and is fully *equal*, if not *superior*, to any other breed. They are the very choicest of the royal stock which is so much admired in England. We are in possession of the shipping papers of these splendid pigs. The freight and incidental expenses on them, alone, amount to about six hundred dollars. They ought to be fine pigs. Three hundred dollars a *pair* for the pigs from this splendid stock would be *low*, taking their great value into consideration. We have often heard of Prince Albert's stock of pigs, but until G. P. Burnham, Esq., of Russet House, Melrose, first imported this superb stock into this country, no American was ever honored with a shy at this extraordinary breed of swine. The company, at great expense and trouble, prevailed upon Mr. Burnham to part with a few of his second-rate samples; and they have now no doubt that they will be able to 'beat him all to rags,' in a few months, since they have been lucky enough to get them from him *pure*ly bred (probably!).

"P. S. Of these pigs, which gained the first prize and gold and silver medal at London in December, 1863, and the first prize and gold and silver medal in Birmingham, were from Tibby, by Wun-eyed Jack. Old Pulgubbin's pigs gained a prize at Mutton-head in 1729, and one at London in 1873."

Still, notwithstanding all this extra flourish of trumpets, the "Porte Monnaie I owe 'em Company" is well-nigh defunct. It was started, unfortunately, about five years and eight months "too late in the season."

Yet, as I honor talent and enterprise, wherever they may be shown, I trust that this association may be galvanized into successful operation — as, *perhaps*, it will!

CHAPTER XXXIX.

A SATISFACTORY PEDIGREE.

In the course of my live-stock experience, and especially during the excitement that prevailed amidst the routine of the hen-trade, I found myself constantly the recipient of scores and hundreds of the most ridiculously unreasonable and meaningless letters, from the fever-struck (and innocent) but uninitiated victims of this epidemic.

In England, amongst other nonsense bearing upon this subject, the more cunning poultry-keepers resorted to the furnishing of *pedigrees* for the birds they sold. This trick worked to admiration in Great Britain for a time, and the highest-sounding names were given to certain favorite fowls, the progeny of which ("with pedigree attached") commanded the most extravagant and ruinous prices, in the English "fancy" market.

For instance, I noticed in the London papers, in 1852, an account given of the sale of "two splendid cinnamon-colored chickens, out of the famous cock 'Jerry,' by the noted hen 'Beauty,' sired by 'Napoleon,' upon the well-known 'Queen Dowager,' grandsire 'Prince Albert,' on 'Victoria First,'" &c. &c., which brought the handsome sum of one hundred and sixty pounds (or about eight hundred dollars). And, soon afterwards, the same dodge was adopted on this side of the Atlantic. The "Porte-Monnaie I owe 'em Company" have *now* an advertisement in several New York and Western papers, concluding thus:

> "To all who desire it, we will furnish authentic pedigrees of our stock of *all* descriptions, which may be relied on for their accuracy."

This sort of thing was rather too much for my naturally republican turn of mind; and, though I could endure *almost* anything in the humbug of this bubble, I could n't swallow *this*. I received from New York State, one day, the following spicy epistle:

"Mr. Burnham.

"Sir: I have been a live-stock breeder for some years in this and the old country, and I was desirous to obtain only *pure*-blooded fowls when I ordered the 'Cochins' of you last month. I asked you for their *pedigree*. You have sent none. What does this mean? I paid you your price — seventy-five dollars — for three chickens. What have you sent me? Am I dealing with a gentleman? Or are you a mere shambles-huckster? What are these fowls bred from? Perhaps I may find myself called upon to speak more plainly, sir. I hope not. Who *are you?* I sent for a pedigree, and I want it. *I must have it*, sir. You will comprehend this, I presume. If you do not, I can enlighten you further. In haste,

"——— — ———."

I smiled at the earnestness of this letter, the more particularly when I reflected that this gentleman always supplied to his patrons a thing he called a pedigree, for all the animals he sold — so intricate, conglomerated and lengthy, that no one would ever venture to dispute the authenticity and reliability of the document he sent them.

I re-read his sharp communication, and I found the sentence again, "Who *are you?* I sent for a pedigree, and I must have it." And I sat down, at once, and wrote him as follows:

"*Melrose, Mass.*, 1853.

"My Dear Sir:

"Your peppery favor came duly to hand. You say you 'want a pedigree,' and that you 'must have it;' and you inquire who *I* am? I cannot furnish any such history for my *fowls*, for I have n't the slightest idea what they are, except that they are bred from my superb imported 'Cochin-Chinas,' which have so long been pronounced the 'admiration of the world.'

"But, since you must have a pedigree, you say, and as you seem anxious to know who *I* am, I enclose you the following, as an accurate account of my *own* pedigree, which I furnished to a legal gentleman in New York city, some years since,* and which, I presume, will answer your purpose as well as any other would; as I observe, by your polite favor now before me, that you 'want A pedigree.' Please read this carefully, and then inform me (as you promise to do) if you 'can enlighten *me* further'!

"Very profoundly yours,

"G. P. B."

It will be necessary, in order that my readers may the better appreciate the pedigree that follows (and which I enclosed to my correspondent, as above stated), to inform them that some fifteen years ago, or more, there was a person named *Burnham*, who died in England, leaving no will behind him; but who was possessed, at the time of his decease, of an immense fortune, said to amount to several millions of pounds sterling in value. As soon as the intelligence reached this country, the Burnhams were greatly elated with their prospects, and meetings of the imaginative

* This article was originally published in the New York *Spirit of the Times*, substantially, and was afterwards issued in an edition of my fugitive literary productions, by Getz & Buck, of Philadelphia, in a volume entitled "*Stray Subjects.*"

"heirs" to this estate were held, who, each and all, believed that a windfall was now in certain prospect before them. The excitement ended as all this sort of thing does. No one among the Burnhams could identify himself, or substantiate the fact of his ever having had a grandfather; and the bubble was soon exploded. Among the parties who were addressed on the subject of this supposed "Burnham fortune," was my humble self; the ambitious lawyer who undertook to unravel the mystery, and to recover the money for us, informing me by mail that "it would be of material pecuniary advantage to me to establish my pedigree." I wrote *him* as follows:

"My Dear Sir:

"Your favor, under date 4th instant, came duly to hand, and I improve my earliest moment of leisure (after the unavoidable delays attendant upon procuring the information you seek) to reply. You are desirous of being made acquainted with my 'pedigree.'

"I have to inform you that I have taken some days to examine into this matter, and, after a careful investigation of the 'records,' find that I am a descendant, in the direct line, from a gentleman, very well remembered in these parts, by the name of Adam. The old man had two sons. 'Cain' and 'Abel' they were called. The latter, by the other's hands, went dead one day; but as no coroner had then been appointed in the county where they resided, 'verdict was postponed.' A third son was born, whom they called 'Seth.' *Cain* Adam had a son named Enoch, who had a son (in the fourth generation) by the name of Malech. Malech had a son whom he called NOAH, from whom I trace directly my own being.

"Noah had three sons, 'Shem,' '*Ham*' and 'Japheth.' The eldest and youngest — Shem and Japheth — were a couple of the 'b'hoys;' and Ham was a very well-disposed young gentleman, who slept at home o' nights. But his two brothers, unfortunately, were not so well inclined. *Ham* was a sort of 'jethro'— the butt of his two brothers, who had done him 'brown' so many times, that they called him 'burnt.' For many years he was known, therefore, as 'Burnt-Ham.' Before his death he applied to the Legislature in his diggings for a change of name. He dropped the *t*, a bill was passed entitling him to the name of BURN-HAM, and hence the *sur*name of your humble servant. So much for the *name.*

"In several of the newspapers of that period I find allusions made to *a very severe rain-storm* which occurred 'just about this time;' and the public prints (of all parties) agree that 'this storm was tremendous,' and that 'an immense amount of damage was done to the shipping and commercial interest.' As this took place some six thousand years back, however, you will not, I presume, expect me to quote the particular details of this circumstance, except in so far as refers directly to my own relatives. I may here add, however, that subsequent accounts inform me that everything of any particular value was totally destroyed. A private letter from Ham, dated at the time, declares that 'there was n't a peg left to hang his hat on.'

"Old Noah found it was 'gittin' werry wet under foot' (to use a familiar expression of his), and he wisely built a canal-boat (of very generous dimensions) for the safety of himself and family. Finding that the rain continued, he enlarged his boat, so that he could carry a very considerable amount of luggage, in case of accident. This foresight in the old gentleman proved most fortunate, and only confirms the established opinion, that the family is 'smart;' for the 'storm continued unabated for forty days and forty nights' (so say the accounts), until every species of animal and vegetable matter had been 'used up,' always excepting the old gentleman's canal-boat *and* cargo.

"Now, Noah was a great lover of animals. 'Of every kind, a male and female,' did he take into his boat with him, and 'a nice time' they must have had of it for six weeks! Notwithstanding the fact (which I find recorded in one of the journals of the day), that 'a gentleman, who was swimming about, and who requested the old man to let him in, upon being refused, declared that he might go to grass with his old canoe, for he did n't think it would be much of

23*

a shower, anyhow,'— I say, notwithstanding this opinion of the gentleman, who is represented as having been a 'very expert swimmer,' everything was destroyed.

"Ham was one of 'em —*he* was ! He 'knew sufficient to get out of the rain,' albeit he was n't thought *very* witty. He took passage with the rest, however, and thus did away with the necessity of a life-preserver. From *Ham* I trace my pedigree directly down through all the grades, to King Solomon, without any difficulty, who, by the way, was reported to have been a little loose in his habits, and was very fond of the ladies and Manzanilla Sherry. He used to sing songs, too, of which 'the least said the soonest mended.' But, on the whole, Sol was a very clever, jolly-good fellow, and on several occasions gave evidence of possessing his share of the cunning natural to our family. Some thought him 'wise;' but, although I have no disposition to abuse any of my ancestors, I think the Queen of Sheba (a very nice young woman she was, too) rather 'come it' over the old fellow !

"By a continuous chain, I trace my relationship thence through a rather tortuous line, from generation to generation, down to Mr. Matthew,— not the comedian, but to Matthew, the Collector (of Galilee, I think), who 'sat at the receipt of customs.' To *this* connection I was, undoubtedly, indebted for an appointment in the Boston Custom-house. Matthew lived in the good old 'high tariff' times, when something in the shape of duties was coming in. But, as nothing is said of his *finale*, I rather think he absquatulated with the funds of the government. But I will come to the information you desire, without further ado.

"You know the 'OLD 'UN,' undoubtedly. (If you don't, there is very little doubt but you will know his *namesake* hereafter, if you don't cease to squander your time in looking after the plunder of the Burnham family!) Well, the 'Old 'Un' is in the 'direct line,' to which I have now endeavored to turn your attention; and I have been called, of late years, the 'YOUNG 'UN,' for reasons that will not interest you. To my honored senior (whom I set down in the category as my legitimate 'dad') I would refer you for further particulars. He is tenacious of the character of his progeny, and loves me; I would commend you to him, for it will warm the cockles of his old heart to learn that the 'YOUNG 'UN' *is in luck*.

"If you chance to live long enough to get as far down in my letter as *this* paragraph, allow me to add that, should you happen to receive any very considerable amount as *my* share of the 'property' for the Burnham family, please not overlook the fact that I am 'one of 'em,' and that I have taken pains to tell you 'whar I cum from.' Please forward my dividend by Adams & Co.'s Express (if their crates should be big enough to convey it), and if it should prove too bulky, turn it into American gold, and charter a steamer to come round for the purpose; I shan't mind the expense.

"In conclusion, I can only intimate the high consideration I entertain towards yourself for having prepaid the postage upon your communication; a very unusual transaction with legal gentlemen. My sensations, upon closing this hasty scrawl, are, I fancy, very nearly akin to those of the Hibernian who '*liked* to have found a sovereign once,' —but you will allow me to assure you that it will afford me the greatest pleasure to meet you hereafter, and I shall be happy to give you any further information in my power touching *that* 'putty' in prospective.

"I am, very respectfully, your obedient servant,

"Geo. P. Burnham, *alias the* 'Young 'Un.'"

I presume this pedigree was perfectly satisfactory to my correspondent; and I am quite certain that it was of as much account as this kind of thing usually is. At any rate, I heard nothing more from him, in any way; and I made up my mind, therefore, that, after reading this, he concluded that he could n't "enlighten me further," as he had so pertly suggested in his communication, quoted in the beginning of this chapter. He is a very nice man, I have no manner of doubt.

CHAPTER XL.

"DOING THE GENTEEL THING."

"THERE is one thing you should always bear in mind," said a notorious shark to me, one day, while we conversed upon the subject of breeding live-stock successfully — "there is one thing you should always remember; and that is, under no circumstances ever permit a fowl or a pig to pass out of your hands to a purchaser, unless you *know* him to be of *pure blood.*"

This is a pretty theory, and, I have no doubt, such a course would work to admiration, if faithfully carried out (as *I* always intended to do, by the way); but in this country this was easier to talk about than to accomplish. I have now a letter before me, received some years since, upon this point, and which will give the reader some idea how far this thing extended in certain quarters, and what came of it.

"SIR: I have been informed by my friends, and I have seen it stated in the poultry-books generally, that *you* are a breeder of fowls who can be relied on. I wish I could say as much of some other parties with whom I have dealt, during the past year or two.

"I have been striving, for a long time, to get possession of some *pure*-bred domestic fowls, and a strain of thorough-bred Suffolk swine. I am satisfied *you* have got them. Now, I beg you will understand that I am fortunately pecuniarily able to *pay* for what I seek. I care nothing for *prices;* * but I do desire, and stipulate for, purity of blood. Can you supply me? What are your strains? When did you import it, and how has it been bred?

"If you can send me half a dozen Chinese fowls, all *pure* bloods, of each of the different varieties, do so, and charge me whatever you please,—only let them be fine, and such as will produce their like.

"I have read much on this subject of poultry, and I want to *begin* right, you perceive. I have made up my mind that there are not so many *varieties* of fowls extant as many breeders describe. I am satisfied that these domestic birds hail originally from China, and that *all* of them are of one blood. What is your opinion?

"Write me your views, please, and let me know if you can furnish me what I seek, upon honor; bearing in mind

* This was the kind of customer I met with occasionally, and whom I always took at his word. The gentleman who "did n't care about price" was always the man after my own heart.

that I am ready to pay your price, whatever it may be; but that I want only pure-blooded stock.

"Yours, respectfully,

"——— ———."

I immediately forwarded to this customer (as I usually did to my newly-found patrons) copies of the *portraits* of my "genuine Suffolk" pigs, and of my "pure-bred" and "imported" Chinese fowls. These "pictures," samples of which appear in this work upon pages 174 and 212, had the desired effect. I rarely forwarded to these beginners one of these nicely-got-up circulars that did n't "knock 'em" at first sight.

These gentlemen stared at the engravings, exclaimed, "*Can* it be?" thrust their hands to the very bottom of their long purses, and ordered the stock by return of mail.

In this last-mentioned case, I informed my correspondent that I agreed with him in the ideas he had advanced precisely (I usually did agree with such gentlemen), and I entertained no doubt that he was entirely correct in his views as to the origin of domestic fowls, of which he evidently knew so much. (This helped me, amazingly.) I pointed out to him the distinction that existed (without a difference) between a "Shanghae" and a "Cochin-China," and finally concluded my learned and *un*selfish appeal by hinting (barely *hinting*) to him that I felt certain *he* was the best judge of the facts in the case, and I would only *suggest* that, so far as my experience went, there were, in reality, but *ten* varieties of *pure*-bred fowls known to ornithologists (I was one of this latter class), and that these ten varieties were the Cochins, the White, Grey, Dominique, Buff, Yellow, Red, Brown, Bronze and Black *Shanghaes*— and these were the only kinds *I* ever bred.

As to their purity of blood, I could only say, that I imported the original stock myself, and "enclosed" he had their *portraits;* to which I referred with pride and confidence and pleasure, &c. &c. &c. Of their probable merits I must leave it entirely to his own good judgment to decide. I had this stock *for sale*, and it did not become me (mind this!) did n't *become me* to praise it, of course (O no!). And I would say no more, but simply refer him to the public prints for my character as a breeder of blooded stock, etc. etc. etc.

Did this take him down? Well, it did; *vide* the following reply from him, two weeks subsequently.

"My Dear Sir:

"I never entertained a doubt that you were *all* you had been represented; and your reputation is, indeed, an enviable one, in the midst of these times, when so much deceit and trickery is being practised among this community. I am flattered with the tone of your kind letter, just received, and I am greatly pleased that you thus readily coincide with me in regard to my opinions touching the fowl race.

"I had come to the conclusion that there were but *eight* real varieties of genuine fowls; but I observe that, in your last favor, you describe *ten* strains of pure-bloods, that you know to be such. The portraits of your stock are beautiful. You allude to the 'Bronze' and the 'Dominique' colored Shanghaes. These must be very fine, I have no doubt; and I gladly embrace the opportunity to enclose you a draft on Merchants' Bank, Boston, for six hundred dollars, in payment for six of each of your splendid varieties of this pure China stock, the like of which (on paper, at least) I have never yet been so fortunate as to meet with.

"Please forward them, as per schedule, in care of Adams & Co.'s Express; whose agents, I am assured, will feed and water them regularly *three times a day** on the route, and who are universally proverbial for their attention to the birds thus directed and intrusted to their care. I shall order the 'Suffolks' shortly. Yours, truly,

"—— ——."

I sent this anxious purchaser sixty chickens, at ten dol-

* Certainly — of course. The express agents had nothing else to do but to "feed and water" fowls "*three times a day*" on the way!

lars each (cheap enough, to be sure), in accordance with his directions, and he was delighted with them. I do not *now* entertain a shadow of doubt that every *one* of those ten "different varieties" were bred from white hens and a black cock, of the ordinary "Shanghae" tribe.

CHAPTER XLI.

THE FATE OF THE "MODEL" SHANGHAES.

NAPOLEON, the great, found himself compelled to succumb to adverse fate, at the end of a long and brilliantly triumphant career. "It was destiny," he said; and he bowed to the fiat, which at last he was unable successfully to dodge.

I was the fortunate owner of a pair of fine Shanghae fowls, that were universally acknowledged to be "at the head of the crowd,"—so far as there was any beauty or attractive qualities, whatever, in this species of animal,—and I thought they were not bad-looking birds, really.

I caused a likeness to be taken of them from life, accurately, and it was placed, some years since, at the head of the circulars which I always enclosed back to my correspondents, in reply to their favors and inquiries regarding my views as to what was the *best* kind of domestic bird for breeding.

The cock was very handsomely formed, and when in full feather was exceedingly showy, and graceful, and noble in his carriage. His hen companions were fine, too; but there was one in particular, that, in company with this bird, I showed at several fairs, where they invariably carried away the first premium, without any question or cavil as to comparative beauty and merit. I named them "Napoleon" and the "Empress."

Their joint weight, when in the best condition, was about twenty-two pounds; and as the "fancy" then raged, they were really unexceptionable. I "donno" how many chickens I have sold by means of the *pictures* of these birds, but I *do* know that, unfortunately, this particular hen never laid an egg while I owned her, which was some two years. Still, she was very handsome, as was also her husband; and I certainly raised a great many fine chickens while they were in my yards. I called them my very best,—and they were, indeed, to look at,—*a model pair of Shanghaes*, as will be seen by a glance at their portraits on the next page.

But they were singled out for a curious fate. At two or three of our early fairs they had taken the first prizes; and at one of the exhibitions, finally, there chanced to come along a gentleman who fancied them exceedingly, and who was bound to possess himself of the best that could be had. He had a long purse (though, at the time *he* bought, prices were not up to the mark they reached subsequently, by a long margin); and when he offered forty dollars for this

THE "MODEL" SHANGHAES.— (See page 280.)

"model" pair, it was thought, by most of the outsiders, to be a fabulous transaction altogether, made up between us to aid in gulling "the people." However, he paid his money for them, sent them off, and the following account of their subsequent fate is thus touchingly furnished by my friend "Acorn," who chanced to be "in at the death":

"The gentleman who became the fortunate purchaser of these fine fowls had come to the city in the morning for the purpose of posting himself up generally, and to procure a pair of these then very desirable birds, though he did not imagine that he would be called upon to come down so 'werry han'some' for a single pair. He saw these, however, and visions of brilliant promise loomed up before him, if he could contrive to obtain them, however high a figure this 'magnificent' twain might be held at. As soon as he secured them, he felt that his fortune was made.

"He calculated to remain in town until evening, and, sitting down, he hastily wrote a note to the keeper of a fashionable hotel in T—— street, informing him that he would dine with him, and that the bearer would deliver him a pair of nice chickens, which he desired him to take charge of. He also directed the boy (to whom he gave this note and the coop) to say that he would take dinner with his friend at four P. M.; and, sending up the fowls, he turned to other matters, for the day.

"Arriving at the hotel, the youngster found the landlord, and said,

"'Here 's a pair of rousing big chickens Mr. M——s has sent up; and he says he 'll be here to dine with you at four o'clock.'

"The landlord supposed that his friend knew a hawk from a handsaw, as well as a canvass-back from a broiled owl; and believed that he had 'sent up' something a little extra for the proposed dinner. He therefore ordered the two birds to be placed in the hands of the cook, and gave directions also to have these 'model Shanghaes' killed and dressed at once, for the proposed dinner, to come off at four o'clock P. M.!

"This order was promptly obeyed; and at the hour appointed the chicken-fancier made his appearance, in company with a few of the 'boys,' and the dinner was served up with due accompaniments. After indulging in sundry wine bitters, as a sharpener to their appetites, the snug party sat down to table, and the liberal owner of the forty-dollar Shanghaes was politely invited to carve. While in the act of dissecting those enormous 'members of the late hen convention,' the amateur remarked,

"''Pon my word, Major, you 've a noble pair of chickens here, to be sure.'

"'Yes, yes,' responded the Major. 'I think they *are* an indifferently good-sized pair of birds. They were sent up to me, to-day, by a mutual friend of ours. I think we shall find them choice.'

"'A present, eh?' said the owner, unwittingly. 'A very clever fellow our friend must be, Major. Capital,—really!' And as he finally commenced to enjoy the feast, he added, 'I declare they are very fine, and of the most delicious flavor I ever tasted. Juicy, too,—juicy as a canvass-back.'

"Thus continued the victim, praising the rich excellence of the birds, until at last he had bagged a bottle or more of sparkling Schreider. While chatting over their Sherry, at last, and enjoying the rich aroma of their regalias, the now unlucky owner of the model Shanghaes suddenly said,

"'By the way, Major, speaking of fowls, what do you think of my hen-purchase, this morning? Are n't they good 'uns?'

"'Well, Bill,' rejoined his friend, 'I think they were delicious. And I won't mind if you dine with me every day in the week, provided you can send me up such chickens as those!'

"'*Such* chickens!' exclaimed Bill, astounded, as the thought for the first time flashed upon him that he might possibly now have been dining upon his 'model Shanghaes.' 'Why, Major, what the deuce do you mean?'

"'Mean?' replied the Major; 'nothing,—only to say —without any intention of disturbing your nerves,—

that we have just finished a most capital dinner upon those nice Shanghaes that you sent up to me, this morning.'

"'*What!*' yelled Bill, jumping wildly up from the table; 'what do you say, Major?'

"'Those Shanghaes —'

"Bill groaned, rammed his hands clean up to the elbows into his breeches-pockets, and, after striding fiercely across the room some half a dozen times, without uttering another word, but with his eyes all this time 'in a fine frenzy rolling,' he stopped short, and, turning to the Major, he exclaimed, with no little gesticulation,

"'Good God, Major, you don't mean to say you're serious, now?'

"'Nothing else, Bill. What's the matter?'

"'Why, *I paid forty dollars for that pair of chickens, this morning, at the hen-show!*'

"'You did!'

"'Yes. Did n't that stupid boy give you my note, when he left the chickens?'

"'Not a note; not even a due-bill,' said the Major, provokingly.

"'I mean my letter,' continued Bill.

"'No,' said the Major, 'he gave me no letter; he simply delivered the fowls, and informed me that you would dine with me at four P. M I thought, of course, you would like them thus, and so I had 'em roasted.'

"Bill did n't stop for further explanations, but rushed for his horse and wagon, and was n't seen in the city but once afterwards, for a long time. He was then closely muffled up, and had both his ears stopped up with cotton-batting, lest he might possibly hear some one say *Shanghae!*

"A few weeks afterwards, while passing near his residence, I halted, and dropped in upon him for an hour; and, after a while, I ventured to touch upon the merits and beauties of the different breeds of poultry; — but I discovered, at once, that there was a wildness about Bill's eyes, and therefore ceased to allude to this usually interesting 'rural' subject, as Bill exclaimed, imploringly,

"'Don't hit me, old boy, now I'm down! *That chicken dinner has never yet digested!*'"

Thus "passed away" one of the handsomest pairs of domestic fowls ever seen in this part of the country, and which were well known, by all the fanciers around me, as tip-top specimens of the then lauded race of Shanghaes.

This result proved rather an expensive dinner for Mr. M——s; but, while it served for an excellent lesson to him (as well as to many of his friends who chanced to hear of what the Major called "this capital joke"), he had the satisfaction, subsequently, of ascertaining that he got off at a remarkably low figure. *His* hen fever was very quickly, and fortunately, cured. But for this sudden and happy turn in his case, the disease *might* have cost him far more dearly.

The fowls he thus lost were what were then deemed "tall specimens;" but they did not, in this respect, equal those of a neighbor, who declared that a young Shanghae cock of his grew so high on the leg, that he got to be afraid of him; and, instead of eating him, one day while the rooster was in a meditative mood, he contrived to place a twenty-feet ladder beside him, and, mounting it, managed to blow out the monster's brains, greatly to the owner's relief.

CHAPTER XLII.

AN EMPHATIC CLINCHER.

ONE of the last specimen *letters* that I will offer I received late in the year of our Lord 1854, which afforded me as much amusement (considering the circumstances of the case) as any one I ever yet received, of the thousands that found their way to "Geo. P. Burnham, Esq., Boston, Mass." Here it is, word for word:

"GEORG BURNAM:

More 'n a yeer aggo i cent yu twenty six dollers in a leter for 3 coshin chiner Chickns, an yu sed tha wus perfeck pure bludds an yu lade yerseff lyble tu a Sute of prosekushn fer letin such dam stuf go intu yure yard or out of it, eether.

"i bred them orl by themselfs an never had no uther cockrill on my plase. an i *no* yu cheeted me like the devl, an yu no it 2. the fust lot of chickns i gut was awl *wite* as snobawls. but i didnt sa nothin, cause wy? Wat did I Want tu let fokes no ide bin fuled an suckt in by a

Corntemtible yanky, fer! i sed nothin an kep shaidy, an stuk to it that i gut em to *breed* wite fouls out on — caus i Ment peeple shudent larf at me, no how!

"Wel, the nex lot of chickns i gut wus *black* as thunder! *black,* Geo Burnam — bred out of yur Patent yaller impoted preemum stock, that yu an the lyin Noospappers ced wus pure bludds. i chocked Every wun on em quicker 'n *scatt* — wen i found um, an ef Yude a bin thare then i guess you Wuddent razed not more'n ten thowsen more fouls to cheet Peeple with after ide a gut a holt on yure desaitful gullet.

"never yu mind now, yuve gut my monny an yu can maik the most of it. aint yu a Pooty kine of mann? dont yu think yu ort tu hav yure Naim put in the nuspapper an let em say more 'n fifty times a Munth that yu breed onny pure Impoted stock? dont yu feel nice wen Yu heer about the luck that peeple has with the stuf you impose on em in this shaimfull maner? Yu muss be a Nise kine of a sort of mann, i *dont* think.

"i tell yu wot i think on yu. i think if yu Shud taik to sum onnest imploiment, sech as drivin a express Waggin or sorring wood, yude be Considurd a gentle mann Compaired with wat yu now be. everybody nose how yu ar cheetin and Gougin and bleadin the publick, an yur naim stinks wuss 'n a ole Hen-cupe enny how. i spose tho ef yu *shud* taik to enny kine of onness sort of way tu git a livin it ud kill yu dam quik cos yu aint uste tu it, an that wud serv yu rite, yu Cheetin lyin onprinsipled nave. ide orter taikn bennits an Minur's advise, an then i Shudent bin suckt by yu. *tha* air Gentle mann to yu, an tha aint no better then tha shud be Neyther — *no* how !

"i dont mine the Eckspence, it aint no cornsidable matter of konsekens Tu me, i 'shure yu. i can *stan* it, yu need n't be Afeered of that. i can aford tu be suckt wunce. But ide like yu tu tell me how Blak chickns an wite chickns an sum of em *orl* Cullers tu, can cum out of pure bludded Aigs, or pure bludded fouls ? tha *carnt,* an yu kno it. an yu kno'de it afore, an yure Welcom tu orl yule evver maik More out of *me,* bait yure life on that, georg Burnam !

"go ahed. suc em as long as Yu can. tha wunt fine yu out fer a wile, an yu can maik sum cornsidable mor Monny out of the flatts, yit. yu thort yude suckt *me* I spoze. well i own up. yu *did.* yu gut twenty six dollers of my monny an i spose yu chukled about it, same's yu did Wen yu stuk them roten aigs onto bill turner. Yude beter cum here, this wa, sum fine da an See the stock here thats bred out of yure preemum fouls. praps Yude git hoam agin without a saw hed. i think yu wood. haddn't yu Better try it on — *hay ?*

"dont yu wish ide pade the postige on this leter ? Yule git a wus wun nex time. ile rite yu agin, wunct a weak, cee ef i dont. ile Meat yu sum day at sum of the *fares* an then cee if i dont Rake yu down with a corse comb. i haint harf dun with yu yit, by a dam site. so wate.

"In haist,

"B—— F—— L——.

"*Poss Skrip.* — P. S. i seen in boston *Times* yisterday that yu 'Lade six aigs on The editurs table, 8 inchis long an 4 inchis Round.' This was put in that paper i Spose sose yu cud cell Aigs. yu ma pool wull over thair ies But yu dont fule *Me.* i doant bleeve yu ever Lade a aig in yur life — yu Hombugg. go tu the devl gorge Burnam!"

A German friend of mine once temporarily left the profession to which he had been educated thoroughly, and, with a few hundred dollars in hand, purchased a small place, a dozen miles out from the city, which was called by the seller of it "a farm."

Mynheer went to work lustily at his new vocation, slaving and sweating and puffing away over his lately acquired grounds, every moment of time that he could borrow or steal from his legitimate duties, and expending upon his "farm" every dollar he could rake and scrape together.

In the fall of his first year as a "practical agriculturist," I met him casually, and I said,

"A——, how does the farming succeed with you? How have you made it?"

"By gar," he replied, "I 'av try vera hard all de time, I 'av plant potato an quash an corn an all dat, I 'av hire all my neighbors to 'elp, I buy all de manoor in town, I 'av spent all my monish—an wot you tink, now, Burnham—wot you tink I get—eh? Well, I git one dam big watermel'n, dass all;—but *he never git ripe, by gar!*"

When I had read the letter which I have just quoted above, I thought of my friend A——, and I said that my correspondent (like a good many before him), as did Mynheer A——, had undertaken a business which was entirely beyond his comprehension.

His letter was complimentary, (!) to say the least of it. But the young man was easily excited, I think. He did pay me some twenty-six dollars for four chickens, and from some cause (unknown to this individual) he got only white or black progeny from the *yellow* fowls I sent him! Was that any business of *mine?* He should have thanked, rather than have abused me, surely,—for did n't he thus obtain a *variety* of "pure" stock, from one and the same source?

Such fortune as this was by no means uncommon. The yellow stock was crossed in China, oftentimes, long before we ever saw it here; and there was only one means of redress that I could ever recommend to these unlucky wights, conscientiously, and that was to buy *more*, and try it again.

Sometimes "like would breed its like" in poultry; not often, however, within *my* humble experience! The amateurs were continually trying experiments, and grumbling, and constantly dodging from one "fancy" kind of fowl to another, in search of the *right* thing; and I endeavored to aid them in their pursuit; though they did not always attain their object, even when they purchased of *me.*

CHAPTER XLIII.

"STAND FROM UNDER!"

I HAVE asserted, in another place, that, in all probability, in *no* bubble, short of the famous "South Sea Expedition," has there ever been so great an amount of money squandered, from first to last, as in the chicken-trade; and, surely, into the meshes of no humbug known to us of the present day have there been so many persons inveigled, as could now be counted among the victims of this inexplicable mania.

A copy of the *Liverpool Express* in January, 1854, now lies before me, from which I notice that the great metropolitan show in London, just then closed, surpassed all its predecessors; and that the excitement in England, at that time, was at its height. The editor asserts that "it was not an easy thing to exhaust the merits of the three thousand specimens of the feathered tribe there shown. No one," continues the writer, "who is at all conversant with natural history, can fail to find abundance of material for an hour's instruction and amusement. The general character of the exhibition has been already indicated; but this is one of those cases in which *no* description, however elaborate, can supply the place of personal inspection."

The British correspondent of the *Boston Post*, but a short time previously, writes that "the fowl fever, which has raged with so much violence in *New England* during the last three years, has extended to this country. There was a great crowing among the cocks at the late Smithfield cattle-show, and there seems to have been a still louder one at the Birmingham fair.

"The mania for the purchase of fine fowls," continues this writer, "was as furious there as if each of them had been the hen in the fable that found the jewel in the dunghill. Some pairs brought as high as forty pounds (two hundred dollars). One very fine pair of Cochin-Chinas sold for fifty pounds (two hundred and fifty dollars). In the catalogue some were marked at one hundred pounds, the *valuation* prices of owners who did not wish to sell. With you, in America, the rage for fowl-raising is simply one of fancy and profit,* but here it is the result — and a very beneficial one, too — of free trade. The price of eggs and poultry, owing to the great demand, does not fall; the price of grain, owing to free importation, does fall; and hence the great profit which is realized from keeping fowls.

---

* *We* have found it a very comfortable "rage," thank you!

The Dorkings are great favorites, less difficult to raise than with you; and, though not abundant layers, still command, from the greater whiteness and superior delicacy of their flesh, a high price in the market. But the new Cochin-China varieties are in the greatest demand; the display of them at Birmingham exceeded all others, and they are now much sought after here."

Such accounts as these continually occupied the papers; and the fever had been kept furiously alive, by this means, until far into the year 1854. The most glowing accounts of the poultry-shows, at home and abroad, were kept up, too; but, in the mean time, Shanghae chickens multiplied rapidly, and grew up, and filled the barns and yards of "the people,"— and at the same time they did not forget how to eat corn, when they could get it.

And, in spite of the best endeavors of interested parties to galvanize the hum into a continued existence, it was now evident to those who watched its progress, as *I* had done, that the death-rattle was clearly in its throat.

At this juncture I was reminded of the details of the mulberry-tree bubble, the tulip fever, and the Merino sheep speculation; and I had taken care not to become involved in the final ruin of the hen-trade (as I knew many had been, and more were destined to be), in the eventual winding-up of this affair, which was now close at hand.

A brief account of the famous sheep mania (so like the

hen fever in its workings) will not be uninteresting at this point; and its record here, perhaps, will have the effect of opening the eyes of some chance reader, haply, who is, even now, half inclined to *try* his hand in the chicken-trade.

This sheep bubble originated in the year 1815 or 1816, immediately after the treaty of Ghent, and at a period when thousands of the American people were actually "wool-mad" in reference to the huge profits that were then apparent, prospectively, in manufacturing enterprises.

In the summer of the last-named year (as nearly as can be fixed upon), a gentleman in Boston first imported some half-dozen sheep from one of the southern provinces of Spain, whose fleeces were of the finest texture, as it was said; and such, undoubtedly, was the fact, though the sheep were so thoroughly and completely imbedded in tar, and every other offensive article, upon their arrival in America, that it would have been very difficult to have proved this statement. But the very offensive appearance of the sheep seemed to imbue them with a mysterious value, that rendered them doubly attractive.

It was contended that the introduction of these sheep into the United States would enable our manufactories, then in their infancy, to produce broadcloths, and other woollen fabrics, of a texture that would compete with England and Europe. Even Mr. Clay was consulted in reference to the sheep; and he at once decided that they were exactly the animals that were wanted; and some of them subsequently found their way to Ashland.

The first Merino sheep sold, if I recollect right, for fifty dollars the head. They cost just *one dollar each* in Andalusia! The speculation was too profitable to stop here; and, before a long period had elapsed, a small fleet sailed on a sheep speculation to the Mediterranean. By the end of the year 1816 there probably were one thousand Merino sheep in the Union, and they had advanced in price to *twelve hundred* dollars the head.

Before the winter of that year had passed away, they sold for fifteen hundred dollars the head; and a lusty and good-looking buck would command two thousand dollars at sight. Of course, the natural Yankee spirit of enterprise, and the love which New Englanders bore to the "almighty dollar," were equal to such an emergency as this, and hundreds of Merino sheep soon accumulated in the Eastern States.

But, in the course of the year 1817, the speculation, in consequence of the surplus importation, began to decline; yet it steadily and rapidly advanced throughout the Western country, while Kentucky, in consequence of the influence of Mr. Clay's opinions, was especially benefited.

In the fall of 1817, what was then deemed a very fine Merino buck and ewe were sold to a gentleman in the Western country for the sum of eight *thousand* dollars; and even that was deemed a very small price for the animals!

They were purchased by a Mr. Samuel Long, a house builder and contractor, who fancied he had by the transaction secured an immense fortune.

Now, Mr. Long had acquired the sheep fever precisely as thousands of others (in later days) have taken the hen fever. And, in this case, the victim was really *rabid* with the Merino mania. In proof of this, the following authentic anecdote will be amply sufficient and convincing.

There resided, at this time, in Lexington, Ky., and but a short distance from Mr. Clay's villa of *Ashland*, a wealthy gentleman, named Samuel Trotter, who was, in fact, the money-king of Kentucky, and who, to a very great extent, at that time, controlled the branch of the Bank of the United States. He had two sheep,—a buck and an ewe,—and Mr. Long was very anxious to possess them.

Mr. Long repeatedly bantered and importuned Mr. Trotter to obtain this pair of sheep from him, but without success; but, one day, the latter said to the former, "If you will build me such a house, on a certain lot of land, as I shall describe, you shall have the Merinos."

"Draw your plans for the buildings," replied Long, instantly, "and let me see them; I will then decide."

The plans were soon after submitted to him, and Long eagerly accepted the proposal, and forthwith engaged in the enterprise. He built for Trotter a four-story brick house, about fifty feet by seventy, on the middle of an acre of land; he finished it in the most approved modern style, and enclosed it with a costly fence; and, finally, handed it over to Trotter, for the *two Merino sheep*. The establishment must have cost, at the very least, fifteen thousand dollars.

But, alas! A long while before this beautiful and costly estate was fully completed, the price of Merinos declined gradually; and six months had not passed away before they would not command twenty dollars each, even in Kentucky.

Mr. Long was subsequently a wiser but a *poorer* man. He held on to this pair till their price reached the par value only of any other sheep; and then he absolutely killed this buck and ewe, made a princely barbecue, called all his friends to the feast, and whilst the "goblet went its giddy rounds," like the ruined Venetian, he thanked God that, at that moment, he was not worth a ducat!

This is absolute, sober *fact*. Mr. Long was completely and irretrievably ruined in his pecuniary affairs; and very soon after this "sumptuous dinner," he took sick, and actually died of a broken heart.

Along in the summer and fall of 1854, having watched the course that matters were taking in the chicken-trade, I became cautious; for I thought I heard in the far-off distance something indefinite, and almost undistinguishable, yet pointed and emphatic in its general tone. I listened;

and, as nearly as I could make the warning out, it sounded like "TAKE CARE!"

And so I waited for the *dénouement* that was yet to come. In the mean time, I had a friend who for five long years had been religiously seeking for that incomprehensible and never-yet-come-at-able *ignis fatuus*, a genuine "Cochin-China" fowl of undoubted purity!

I had not heard of or from him for some weeks; until, one morning, about this time, a near relative of his sent to my house all that remained of this indefatigable searcher after truth; an accurate drawing of which I instantly caused to be made — and here it is!

CHAPTER XLIV.

BURSTING OF THE BUBBLE.

My friend John Giles, of Woodstock, Conn., has somewhere said, of late, "I often hear that the 'fowl' fever is dying out. If by this is meant the unhealthy excitement which we have had for a few years past, for one, I say the sooner that it dies out the better. But as to the enthusiasm of *true* lovers of the feathered tribe dying out, it never will, as long as man exists. It is part of God's creation. The thinking man loves and admires his Maker's work; always did; always will. And I have not the least doubt that any enterprising young man, with a suitable place and fancier's eye, would find it to his advantage to embark in the enterprise of fowl-raising for market."

Now, I don't know but John is honest in this assertion,—that is, I can imagine that he believes in this theory! But how he can ever have arrived at such a conclusion (with the results of his own experience before him), is more than I *can* comprehend.

Laying aside all badinage, for the moment, I think it may

be presumed that I have had some share of experience in this business, *practically*, and I think I can speak advisedly on this subject. As far back as during the years 1839, '40 and '41, I erected, in Roxbury, a poultry establishment on a large scale, upon a good location, where I had the advantages of ample space, twenty separate hen-houses, running water and a fine pond on the premises, glass-houses (cold, and artificially heated, for winter use), and every appurtenance, needful or ornamental, was at my command.

I purchased and bred all kinds of domestic fowls there, and they were attended with care from year's end to year's end. But there was *no* profit whatever resulting from the undertaking,— and why?

The very week that a *mass* of poultry — say three to five hundred fowls — is put together *upon one spot*, they begin to suffer, and fail, and retrograde, and die. No amount of care, cleanliness or watching, can evade this result. *In a body* (over a dozen to twenty together), they cannot thrive; nor can the owner coax or force them to lay eggs, by any known process.*

* Since this was written, I find in the *Country Gentleman* a communication from L. F. Allen, Esq., on this very subject, in which he says that "A correspondent desires to know how to build a chicken-house for 'about one thousand fowls.' If my poor opinion is worth anything, *he will not build it at all.* Fowls, in any large number, will not thrive. Although I

To succeed with the breeding of poultry, the stock must be *colonized* (if a large number of fowls be kept), or else only a few must find shelter in any one place, about the farm or country residence. And my experience has taught me that five hens together will yield more eggs than fifty-five together will in the same number of months.

I honestly assert, to-day, that of all the humbug that exists, or which has been made to exist, on this subject, no part of it is more glaringly deceptive, in my estimation, than that which contends for the *profit* that is to be gained *by breeding poultry — as a business by itself — for market consumption.* The idea is preposterous and ridiculous, and no man can accomplish it,— I care not *what* his facilities may be,— to any great extent, *upon a single estate.* The thing is impossible; and I state this, candidly, after many years of practical experience among poultry, on a liberal scale, and in the possession of rare advantages for repeated experiment.

I do not say that certain persons who have kept a *few* fowls (from twenty-five to a hundred, perhaps), and who have looked after them carefully, may not have realized a profit upon them, in connection with the farm. But, to

have seen it tried, I never knew a large collection of several hundred fowls succeed *in a confined place.* I have known sundry of these enterprises tried; but I never knew one *permanently* successful. They were all, in turn, abandoned." The thing is entirely impracticable.

make it a business *by itself*, I repeat it, a *mass* of domestic and aquatic fowls cannot be kept together to any advantage whatever, their produce to be disposed of at ordinary market value.

The fever for the "fancy" stock broke out at a time when money was plenty, and when there was no other speculation rife in which every one, almost, could easily participate. The prices for fowls increased with astonishing rapidity. The whole community rushed into the breeding of poultry, without the slightest consideration, and the mania was by no means confined to any particular class of individuals — though there was not a little shyness among certain circles who were attacked at first; but this feeling soon gave way, and our first men, at home and abroad, were soon deeply and riotously engaged in the subject of henology.

Meantime, in England they were doing up the matter somewhat more earnestly than with us on this side of the water. To show how even the nobility never "put their hand to the plough and look back" when anything in this line is to come off, and the better to prove how fully the poultry interests were looked after in England, I would point to the names of those who, from 1849 to 1855, patronized the London and Birmingham associations for the improvement of domestic poultry.

The Great Annual Show, at Bingley Hall, was got up under the sanction of His Royal Highness Prince Albert, the Duchess of Sutherland, Lady Charlotte Gough, the Countess of Bradford, Rt. Hon. Countess of Littlefield, Lady Chetwynd, Hon. Viscountess Hill, Lady Littleton, Hon. Mrs. Percy, Lady Scott, and a host of other noble and royal lords and ladies, whose names are well known among the lines of English aristocracy.

But, as time advanced, the star of Shanghae-ism began to wane. The nobility tired of the excitement, and the people of England and of the United States began to ascertain that there was absolutely nothing in this "hum," save what the "importers and breeders" had made, through the influence of the newspapers; and while a few of the *last men* were examining the thickness of the shell, cautiously and warily, the long-inflated bubble burst! and, as the fragments descended upon the devoted heads of the unlucky star-gazers, a cry was faintly heard, from beneath the ruins — "*Stand from under!*"

I had been watching for this climax for several months; and when the explosion occurred, as nearly as I can "cal-'late," *I* was n't *thar!*

CHAPTER XLV.

THE DEAD AND WOUNDED.

I HAVE never yet been able to ascertain, authentically, all the exact particulars of the final catastrophe; but, basing an opinion upon the numerous "dispatches" I received from November, 1854, to February, 1855, the number of dead and wounded must have been considerable, if not more. I received scores of letters, during this last period mentioned, of which the annexed is a fair sample:

"G. P. BURNHAM, ESQ.

"DEAR SIR: I'm afraid the jig is up! There's a big hole in the bottom somewhere, or I am mistaken. I *think* the dance is concluded; and if it is n't time to 'blow out the lights' and shut down the gate, just let us know,—will you? Where's Bennett, and Harry Williams, and Dr. Eben, and Childs, and Ad. White, and Brackett, and Johnny Giles, and Uncle Alden, and Buckminster, and Chickering, and Coffin, and Fussell, and Chenery, and Gilman, and Hatch, and Jaques, and Barnum, and Southwick, and Packard, and Balch, and

Morton, and Plarsted, and Geo. White, *et id omne genus?* Where are they all? *S-a-y!*

"What has become of Platt, and Miner, and Newell, and Hudson, and Heffron, and Taggard, and Hill, and Swett, and M'Clintock, and Dr. Kerr, and Devereux, and Thacher, and Haines, and Hildreth, and Brown, and Smith, and Green, and *their* allies? Are they dead, or only 'kilt'? Let me know, if you can, I beseech you!

"'O, where, tell me where,' is my bonnie friend John Moore, and mine ancient *frère* Morse, and my loved chum Howard, and the wily Butters? And where's Pedder — the immaculate Pedder? And Charley Belcher, too, and bragging Cornish, and Billy Everett, and our good neighbors Parkinson, and George, and Sol. Jewett, and President Kimball, and know-nothing King, and the reverend Marsh, and Pendletonian Pendleton of Pendleton Hill, and their satellites? Have *all* departed, and left no *wreck* behind? I reckon not!

"Seriously, friend B——, what does all this mean? Has the fever passed by? Can't we offer another single prescription? Has the *last* man been heard from? Has there been found 'a balm in Gilead' to heal the wounds of the afflicted sufferers? Is the thing finished? Are they all cured? Did you say *all?* Dunder and blixen! Is anybody hurt? What are we to do? '*Speak*, or die!'

"Where are the 'Committee,' and the 'Judges,' and the

'Trustees,' and the 'Managers'? Where is the 'Society' whose name, 'like linked sweetness long drawn out,' I have n't time to write? Where is *that* balance in the *Treasurer's* hands,'—and where is that functionary himself? Did he ever exist at all? What has become of the premiums that were *awarded* at the last show in Boston? And when, in the language of the enthusiastic Mr. Snooks (at the Statehouse in 1850), will that Association begin 'to be forever perpetuated,'— eh?

"I have got on hand three hundred of the Shanghae devils! What can I do with them? There is a neighbor of mine (a police-officer), who has got stuck with a lot of 'Cochin' chickens, which he swears he won't support this winter; and he has at last advertised them as *stolen property*, in the faint hope, I suppose, that some 'green 'un' will come forward and claim them. You can't get rid of these birds! It is useless to try to sell them; *you can't give them away;* nobody will take them. You can't starve them, for they are fierce and dangerous when aggravated, and will kick down the strongest store-closet door; and you can't kill them, for they are tough as rhinoceroses, and tenacious of life as cats. Ah! Burnham, I have never forgiven the man who made me a present of my first lot! Do you want what I've got left? Will you take them? How much shall I pay you to receive them? Help me out, if you can.

"I am not aware that I ever committed any offence, that this judgment should be thus visited upon *my* poor head! I never sold fowls for what they were *not;* I never cheated anybody, that I know of; I do not remember ever having done any unjust act that should bring down upon me this terrible vengeance. Yet I am now the owner of nearly three hundred of these infernal, cursed, miserable ghosts in 'feathered mail,' which I cannot get rid of! Tell me what I shall do, and answer promptly.

"Yours, in distress,

"—— — ——."

I have smiled over this document, so full of feeling and earnestness, so lively and touching in its recollections of the days when we went *chicken*-ing, long time ago! But I have never been able to reply fully to my ardent friend's numerous inquiries. I don't want those "three hundred Shanghae devils," though. I have now on hand *nine* of them (only, thank Heaven!) myself; and that is quite enough for one farm, at the present current price of grain.

What has become of all the friends about whom my correspondent so carefully inquires, I don't know. Not *five* of them are now *in* the hen-trade, however; and there are not ten of them who got *out* of the business with a whole skin, from the commencement.

The engine has collapsed its boiler. There was altogether too much steam crowded on, and the managers were not all "up to snuff." The dead and wounded and dying are now scattered throughout New England and New York State chiefly, and their moans can occasionally be heard, though their groans of repentance come too late to help them.

They recklessly invested their twenties, or fifties, or hundreds, and, in some instances, their thousands of dollars, in this hum, without any knowledge of the business, and without any consideration whatever, except the single aim to keep the bubble floating aloft until they could realize anticipated fortunes, on a larger or smaller scale, as the case might be. But the "cars have gone by," and they may now wait for another train. *Perhaps* it will come!

Poor fellows! Poor, deluded, crazy, reckless dupes! You have had your fun, many of you, and you will now have the opportunity to reflect over the ruins that are piled up around you; while, for the time being, you may well exclaim, with the sulky and flunkey Moor,

"*Othello's occupation 's* GONE!"

CHAPTER XLVI.

A MOURNFUL PROCESSION.

I WAS sitting before my comfortable library fire in mid-winter, 1854, and had been reflecting upon the mutability of human affairs generally, and the uncertainty of Shanghae-ism more particularly, when I finally dropped into a gentle slumber in my easy-chair, where I dozed away an hour, and dreamed.

My thoughts took a very curious turn. I fancied myself sitting before a large window that opened into a broad public street, in which I suddenly discovered a multitude of people moving actively about; and I thought it was some gala-day in the city, for the throng appeared to be excited and anxious. "The people" were evidently abroad; and the crowds finally packed themselves along the sidewalks, leaving the wide street open and clear; and I could overhear the words "They 're coming!" "Here they are!"

I looked out, and beheld an immense gathering of human beings approaching in a line that stretched away as far as the eye could reach,—a dense mass of moving mortality, that soon arrived, and passed the window, beneath me. I was alone in the room, and could ask no questions. I could only see what occurred before me; and I noted down, as they passed by, this motley PROCESSION, which moved in the following

Order of March.

ESCORT OF INDESCRIBABLES.

Hatless Aid. [Chief Marshal in Black.] Bootless Aid.

Police. Two EX-MORMONS IN WHITE TUNICS. Police.

Calathumpian Band.

Whig Office-holders. { The "*Know Nothing Guards*," with guns enough for all useful purposes. } Democrat Expectants.

THE "INS." [COLLECTOR and POSTMASTER.] THE "OUTS."

U. S. Marshal. { The "National" Democracy, two deep, in one section. } U. S. Dist. Att'y.

BANNER.

Motto — "*We know of* BURNS *that Russia Salve can't cure.*"

"Aids to the Revenue." [MARSHAL.] Drawbacks on the Revenue.

Kaleb Krushing. [THE MAN WHO FAINTED IN MEXICO.] Jorge ah ! Poll.

"*Fanny Fern,*"

Flanked by a company of disappointed Publishers, twenty-four deep, in twelve sections.

BANNER.

Motto. — "*She 's a brick !*"

Aids. [MARSHAL.] Aids.

President of the "N. E. Mutual Admiration" Hen Society.

Fat Marshal. [THE GREAT SHOW MAN.] Lean Marshal.

BAND, playing the "Rogue's March."

Marshal. GHOST OF JOICE HETH. Marshal.

Aids, of Quaking Shakers. { A Fejee Mermaid, astride the Woolly Horse. } Aids, The Happy Family.

Aids, Their readers { *Invited Guests.* *The Three Historians,* BURNHAM, PRESCOT, and BANCRAFT. } and "admirers,"

Escort in the rear, with charged bayonets.

Police. { A *genuine* "COCHIN-CHINA" Rooster, succeeded by the man who *knew* him to be such ! } Police.

Marshal. Pea Wilder. { The entire United States American National Agricultural Society, in a one-horse buggy. } Marshal. W. ESS king.

[The *good* this association had accomplished was borne along by a stout "practical farmer," in a small thimble; the *records* of its doings were inscribed on a huge roll of paper, 16,000 yards long, carried upon a truck drawn by twelve yoke of "pure" Devon oxen.]

BANNER. — Motto: "*Ourselves and those who vote for us.*"

Aid, Naval Store Keeper.	An ex-U. S. Navy Agent who left that office without having made money out of his place! BANNER. — Motto: "*Poor, but honest.*"	Aid, U. S. Sub-Treasurer.
One hundred and twenty-five Marshals.	The Mass. Hort. Improvement Society, *en masse*, with several full bands of music, on "seedling" accompaniments, etc. BANNER. Motto: "*Cuss the Concord Grape.*"	Twenty-five hundred and one gold-medal seekers.
No Aids.	The man who voluntarily gave up his office under the National government, *solus*, on horseback, with BANNER. — Motto: "*Few die, and none resign.*"	No friends.
The defunct New England Hen Society.	"THE YOUNG 'UN," in his own barouche, drawn by four "superb dapple-grey Shanghaes." BANNER. — Motto: "*Who's afraid?*" MUSIC. BANNER. — Motto: "*Not this child!*"	His vanquished Competitors.
Police.	HEN MEN WHO HAD MISTAKEN THEIR CALLING, twenty-eight deep, in four hundred sections.	Police.
Aids, 24 Constables.	GRAIN MEN, *with their bills,* in seventy sections, sixty-four deep.	Aids, All in a row.

BAND, playing "*Hope told a flattering tale.*."

Tree-venders and Horticulturists,	The great-grandson of the man who set out an orchard of dwarf Pear-trees (in a barouche). He was 102 years old, and believed he should see fruit on those same trees "next season"!	with thumbs on their noses.
Pall Bearers.	[HIS COFFIN, BEHIND.]	Heirs to his estate.
Aids, 12 Respectable Physicians.	BELIEVERS THAT COCHITUATE WATER IS WHOLESOME (in a chaise).	Aids, Board of Commissioners.
15 Marshals.	Chicken Fanciers who *did n't* buy their eggs of me, and who *expected they would hatch!* (Four thousand strong.)	15 Marshals.
Aids, the Conductors.	A body of Express Agents, who never shook up the eggs intrusted to them (though they occasionally shook down their employers).	Aids, the Brakemen,

BAND. — Air: "*O, I never will deceive you!*"

Flanked by the Subscribers for *that* "Double Harness,"	"*My friend* THE PRESIDENT," In the carriage presented to him by "the people," drawn by that "*superb* pair of $1500 horses" which we read of in the papers.	and the "mourners" who *did n't* obtain fat offices.
Motto: "*I'll see you in the Fall.*"	BANNERS.	Motto: "*Save me from my friends!*"
	Full Band.	
Aid, BRASS & CO.	The Hatch Grey Shanghae Express Co., with the latest news from Nantucket and "Marm Hackett's Garden." *Motto:* "IMPORTANT, IF TRUE!"	Aid, THE "COLONEL."
Aid, Two Presidents.	Holders of *Second* Mortgage R. R. Bonds, 24 deep, in 2400 sections. BANNER. Motto: "*There 's a good time coming.*"	Aid, One Treasurer.
Aids, 5 Regular Doctors.	The owner of the first "Brahma Pootra" fowls in America, with a map of India on the seat of his pantaloons.	Aids, Faculty of Ripum College.
Aid, Lucy Brick.	The original members of the "WOMEN'S RIGHTS CONVENTION." BAND. — Air: "*Why don't the men propose?*"	Aid, Abby Fulsome.
Aids, The First Premium Fowls.	The "wreck of Burnham's character" caused by the *powerful* newspaper assaults of one TEE BEE MINUR, A.SS.	Aids, The "Porte-Monnaie I owe 'em Company."
	BANNER. — Motto: "*Don't* he feel bad!"	
No aid for him!	The Poultry Fancier who had found out the exact difference between a "Cochin-China" and a "Shanghae."	Too far gone!
Unpaid Compositors.	Delinquent subscribers to northern *Farmers*, twelve deep, and three miles long!	Disappointed "Press Gang."
Marshal.	[The "editor," suffering from a severe attack of *roup.*]	Marshal.
DAVID.	Dr. Bangit, with the unsold copies of his Poultry-Book, in a huge baggage-wagon, drawn by 14 horses.	GOLIAH.
Aids, 15 Sisters of Charity.	A battalion of victims to the Hen Fever, who had bought eggs that "did n't hatch," and who were waiting patiently to have their money returned!	Aids, 15 friends to the Insane Poor.
Marshal and Deputy Sheriff.	My *legal* friend (on a mule) who promised to spend a thousand dollars in prosecuting me for selling him *Shanghae* eggs for *Cochin-Chinas!*	Jail Keeper and 4 Constables.

Left	Procession	Right
Aid, Barnam.	Fat Johnny Jiles, with the head of a *pure* "Black Spanish" crower on a salver.	Aid, Burnum.
Marshals.	The men who *did n't* take the first premiums (when I was round) at the Poultry-Shows (in deep mourning).	Marshals.
Aids, A "Cabinet" of Curiosities.	The political remains of Frank Pierce, in a toy wheelbarrow, with BANNER, on a "sharp stick." Motto: "*Veto.*"	Aids, His own Opinions!
Aid, Editor of the *Northum Farmer.*	Victims who purchased Minor's "Patent Cross-grained Collateral Beehives," with Motto: "*Burned child dreads the fire.*"	Aid, GEN. BANGIT, of the "Nauvoo Legion."
Aids, The Sellers.	*Customers for "Ozier Willow,"* in two sections, *one man deep.* BANNER. — Motto: "*I rather guess not!*"	Aids, The Victims.
Marshals.	A huge concourse of "Copper Stock" and "Agewuth Land" owners, in deep sables.	Marshals.

FULL BAND. — Air: *Dead March.*

BANNER. — Motto: "*You 're sure to win — if you don't lose!*"

☞ A smooth-skinned *pure* "Suffolk" Pig, *imported.* ☜

Twenty-four Sewing Machines, "*warranted.*"

Left	Procession	Right
Aid, Secretary.	President of the "PORTE-MONNAIE I OWE 'EM COMPANY," as Richard III. on horseback.	Aid, Treasurer.

Nine "Bother'em Pootrums," rampant.

Left	Procession	Right
The few unlucky Buyers.	The identical lot of "pure-bred" fowls that Bangit, Plarsterd, Minor, Humm & Co., *imported* (over the left) "for the Southern market," in 1853!	The Believers in this story!

The Hen that lays two eggs a day!

Treasurer of the "Mut. Adm'n Society."

Defunct Hucksters, in a tip-cart.

Four empty Hen-Coops, on wheels.

☞ Breeders of *pure* Alderney cattle! ☜

who furnish Pedigrees with long tails.

An effigy of the LAST MAN that will buy Shanghae chickens (in a strait-jacket).

Left	Procession	Right
Police and Aids.	Purchasers of Live Stock who bought of my competitors; with BANNER. Motto: "*We got more than we bargained for!*"	Sheriff and posse.

The Hen-Men who "pity Poor Burnham."

MY OWN CASH CUSTOMERS,

10,000 *strong!*

CAVALCADE.

"THE PEOPLE,"

MUSIC,

And the rest of Mankind,

etc.

The scene was closing! That immense concourse of humbugs and humbugged had passed on, and I was alone once more. But, a moment afterwards, I saw the head and face of a comical and good-humored looking Yankee (just beneath the window), who was in the act of puffing into the air a huge budget of bubbles, that danced and floated in the atmosphere for a brief moment, and which, bursting, suddenly awoke me. Here is a sketch of the *finale.*

CHAPTER XLVII.

MY SHANGHAE DINNER.

I SAW by the papers, one day, late in the year 1854, an account of the return from England of my fat friend Giles, who brought with him the poultry purchased abroad for Mr. Barnum, and which proved to be a lot of *pure* stock, of a remarkable character, as I supposed it would be.

But, while John was absent in Great Britain, the knowing ones there shook him down, beautifully! His theory, when he left America, four months previously, was that "hall 'at was wanted 'ere was to get hover from Hingland pure-bred fowls, and such would *sell.*" John brought over "such," and they *did* sell; but Barnum was sold by far the worst!

An auction was immediately got up at the American Museum, in New York; and after a vast deal of drumming, puffing and advertising this magnificent, just-imported, pure-bred poultry, the sale came off, to a sorry company, indeed! And the gross amount of the sales of the fowls thus disposed of, really, was insufficient to pay the *freight* bills for bringing them across the Atlantic, to say nothing of their original high cost abroad. The show-man has since left the hen-business, I learn, "a wiser if not a better man;" while John retired with the simple exclamation, "Most extr'ornerry result I hever 'eer'd of in hall my life!"

Soon after this little episode occurred, the second show of the "National Poultry Society" (in January, 1855) came off at Barnum's Museum, in New York; which, notwithstanding the best endeavors of the "President," was a failure. The "Committee" shut out of their premium list the Grey Shanghaes, altogether; and the result of this last exhibition was just what I had anticipated. But Mr. Barnum can well afford to foot the bills; and, as he is perfectly willing to do this, no objection will be raised to his choice, I presume. This final exhibition at New York, I have no doubt, closed up the business, for the present.

As soon as this last fair had closed, and when the lucky and unlucky contributors returned to Boston, I invited a party of my former *confrères* to my residence, to dinner. I had been preparing for this little event for several days; and the following was the actual "bill of fare" to which we all sat down, at Russet House, *Melrose*, on the fifth day of February, 1855:

Dinner.

SOUP—*A la* SHANGHAE.

FRESH FISH—WITH CHINA SAUCE.

BOILED FOWL—*To wit, the identical Grey Shanghae cock (two years old) which took the premium at the* FIRST *National Poultry Show, in New York, in* 1854; *then valued at* $100.

ROAST—Shanghae Cock, nine months old, weighing, dressed, 10¾ pounds.
Do. Shanghae Pullets, same age, drawing, dressed, 7½ pounds each.
Do. Spring Shanghae Chickens in variety.

BAKED—Pure "Suffolk" Pig, with genu-wine "Mandarin" Sauce.

ENTREMENTS.

BROILED Shanghae Chicks.
STEWED Shanghae Chickens.
CURRIED Shanghae Fowls.
FRIED Shanghae Pullets.
CODDLED Shanghae Stags.
FRICASSEED Shanghaes.

SHANGHAES TRUFFLED,
and
MORE SHANGHAES, IF WANTED!

DESSERT.

Shanghae Chicken Pie.
Shanghae Omelets.
Shanghae Custards.
Chinese Pudding.
Pudding *a la* Shanghae.
Candied Cocks' Spurs.
Crystallized Pullets' Combs.
Shanghae Wattles, in Syrup.

SHANGHAE-QUILL TOOTH-PICKS,
and
MORE SHANGHAES IN THE YARD!

To this repast, with thankful hearts, a company of five-and-twenty sat down, and, as nearly as my recollection now serves me, the friends did ample justice to my Shanghae dinner. After two hours over the *varied* dishes (varied in size and style of cooking only), the cloth was removed, and the intellectual treat commenced with a song, written "expressly for this occasion," by the Young 'Un, which was delivered with admirable effect by "one who had been there," and in the chorus of which the guests unitedly joined, with surprising harmony and unison. The following toasts were then submitted:

By the Man in the Black Coat. — The Memory of the defunct Rooster we have this day devoured: Peace to his manes! (Drank standing, in silence.)

By a Successful Breeder. — The health, long life, and prosperity, of our absent cash customers, — at home and abroad.

By an Amateur. — Honor to the discoverer of the exact difference between a "Shanghae" and a "Cochin-China" fowl, if he shall ever turn up!

By the "Confidence" Man. — The Continuity of the beautifully-elongated Chinese fowls: May their shadows never be less!

By a Victim. — The Bother'em Wot-yer-call-'ems: Dammum! (Nine cheers for Doctors Bennett and Miner.)

By a Disappointed " Fancier." — Barn-yard fowls and white-shelled eggs, for *my* money. (Three cheers for the old-style biddies.)

By the Youth in a White Vest.—"Fanny Fern": The hen that lays the golden eggs. (Six cheers for Fanny, and the fair sex generally.)

By a Repentant. — The whole Shanghae Tribe: Curse 'em; the more *fowls* you see of this race, the less *eggs* there are about! (This was deemed slightly *personal*, but it was permitted to pass; the gentleman spoke with unusual feeling; he had been only three years in the trade, and had expended some sixteen hundred dollars in experimenting with a view to establish a *breed* that would lay *two* eggs daily.)

By One of my " Friends." — The Young 'Un: The only hen-man who has put the knife in up to the handle with a decent grace! (Nine cheers followed, for the importer of the only pure-bred poultry in America.)

This last sentiment called me to my feet, naturally enough; and, as nearly as I remember, I thus addressed my guests, amidst the most marked and respectful attention:

"Gentlemen: I think I have seen it written somewhere, or I have heard it said, 'It is a long lane that has no turn in it.' I believe, however, that, although the lane we have most of us been travelling for the last six years has proved somewhat tortuous as well as lengthy, we have now passed the *turn* in it, and have arrived very nearly at the end of the road.

"Few of you, gentlemen, have met with so many thorns, *en route*, as I have; none of you, perhaps, have gathered so many roses. I am content, and I trust that everybody is as well satisfied with the results of this journey as *I* am. *The Shanghae trade is done*, gentlemen! We have this day eaten up what, four years ago, would have been the nucleus, at least, of a small fortune to any one of us who at that time might have chanced to have possessed it. But the fever is over; the demand for giraffe cocks and chaise-top hens is passed; the 'poor remains of beauty once admired, in my premium fowls,' now lie scattered about the dishes that have just left this table; and 'Brahma-pootra-ism' is now no longer rampant.

"Perhaps, gentlemen, as you entertain opinions of your own upon this delightfully pleasing subject of poultry-raising generally, and of the propagation of Shanghae fowls in particular, you would care to hear nothing of my views regarding this point. Yet, I pray you, indulge me for a single moment — in all seriousness — and permit me to say (without the slightest intention of being personal), that we have proved ourselves a clan of short-sighted mortals, at the best, during the last half-dozen years, in our crazy devotion to what we have deemed an honorable and laudable 'profession,' but which has been, in reality, the most shallow, heartless, unreasonable, silly and bottomless humbug that grown-up men have ever been cajoled with, since the hour when Adam was fooled by the accomplished and coquetting Eve!" (Cries of "You're more 'n half right!" "That's a fact!" "Exactly — just so!")

"There is now living in Melrose, Mass., gentlemen, a breeder who begun at the beginning of this excitement, who has since followed up the details of this hum with a zeal worthy of a better cause; and who has accumulated a handsome competency in this traffic, by attending strictly to his own affairs, while he has uniformly acted upon the principle that this world is sufficiently capacious to accommodate all God's creatures, without jostling. If you should chance to meet this now retired fowl-fancier, he will tell you that he has had, and believes he still has, many personal friends; but the very *best* 'friend' he has ever known is the enjoyment of his present income of eight per cent. interest, per annum, upon thirty thousand dollars. But this is a digression, and I beg pardon for the allusion.

"I look back with no regrets at the past, gentlemen. We have seen a great many merry days, and, in the midst of the competition and humbuggery in which we enlisted, we have often differed in sentiment. But *here*, — at the close of the route on which we have so long been journeying, — let us remember only the good traits that we any of us possess, while from this point we forget the errors that ourselves and our companions may have committed, forever." (Three times three, "and one more," were here given for the speaker, his friends, and all the rest of mankind.)

"I will say no more, gentlemen. My stomach is too full for further convenient utterance; and I will conclude with a sentiment to which, I am sure, you will all respond. I will give you —

"'The Hen Fever!'" —

"Don't, *don't!*" shrieked the crowd. "We 've had that disease once, and that is quite sufficient."

"Indulge me, gentlemen, one moment, and I will propose, then —

"'THE HEN TRADE: Though a *fowl* calling, it puts *fair* money in the purse, when "judiciously" managed. May none of you ever do worse, pecuniarily, in this humble "profession," than has your friend — the subscriber.'"

Another round of hearty cheers succeeded this sentiment, a parting bumper was enjoyed, and the circle separated, to meet again at Philippi, — or elsewhere, — where the author hopes to encounter only friendly faces, whatever may have been his business relations with his acquaintances in the days that are now passed away.

The mania is over. I have frankly repeated to you the humble history of this curious fever, and we have reached

THE END.

LIST OF BOOKS

PUBLISHED BY

JAMES FRENCH & CO.,

78 Washington Street, Boston.

SCHOOL BOOKS.

FOSTER'S BOOK-KEEPING, BY DOUBLE AND SINGLE ENTRY, both in single and copartnership business, exemplified in three sets of books. Twelfth Edition. 8vo. Cloth, extra. . 1 00

FOSTER'S BOOK-KEEPING, BY SINGLE ENTRY, exemplified in two sets of books. Boards. 38

FRENCH'S SYSTEM OF PRACTICAL PENMANSHIP, founded on scientific movements ; combining the principles on which the method of teaching is based. — Illustrated by engraved copies, for the use of Teachers and Learners. Twenty-seventh Edition. 25

This little treatise seems well fitted to teach everything which can be taught of the theory of Penmanship. The style proposed is very simple. The copperplate fac-similes of Mr. French's writing are as neat as anything of the kind we ever saw. — *Post.*

Mr. French has illustrated his theory with some of the most elegant specimens of execution, which prove him master of his science. — *Courier.*

This work is of a useful character, evidently illustrating an excellent system. We have already spoken of it in terms of approbation. — *Journal.*

This little work of his is one of the best and most useful publications of the kind that we have seen. — *Transcript.*

BEAUTIES OF WRITING, containing twenty large specimens of Ornamental Penmanship, Pen Drawing, and off-hand Flourishing. 75

BOSTON COPY-BOOK; comprising nearly two hundred engraved copies, for the use of Schools and Academies. . . 42

LADIES' COPY-BOOK, containing many beautiful engraved copies, which are a perfect imitation of the natural hand writing; also including German Text and Old English. . . 17

BOSTON ELEMENTARY COPY-BOOK, comprising large and small Text Hand, for Schools. 12½

COOK'S MERCANTILE SYSTEM OF PENMANSHIP. Fourth Revised Edition. 37½

THE ART of PEN-DRAWING, containing examples of the usual styles, adorned with a variety of Figures and Flourishes, executed by command of hand. Also a variety of Ornamental Penmanship. 75

MISCELLANEOUS AND JUVENILE.

TURKEY AND THE TURKS, by Dr. J. V. C. Smith, Mayor of Boston. 320 pages. 12mo. Cloth. 75

It is a most excellent work. It will have a large sale, for it embraces more real information about real Turks and their strange peculiarities than anything we have yet read. — *Post.*

THE MASSACHUSETTS STATE RECORD, for the years 1847, 1848, 1849, 1850 and 1851; one of the most valuable American Statistical Works. 5 vols. 12mo. Cloth. . . 5 00

THE NEW HAMPSHIRE FESTIVAL. A graphic account of the Assemblage of the "Sons of New Hampshire" at Boston, Hon. Daniel Webster presiding. Illustrated with portraits of Webster, Woodbury and Wilder. 8vo. Cloth, gilt. . 2 00
THE SAME, Gilt Edges and Sides. 3 00

SECOND FESTIVAL of the "Sons of New Hampshire." Illustrated with portraits of Webster, Wilder, Appleton and Chickering. 8vo. Cloth, gilt. 2 00
THE SAME, Gilt Edges and Sides. 3 00

FESTIVAL. 2 vols. in one. 8vo. Cloth, gilt. . . 2 50

ELEANOR: OR, LIFE WITHOUT LOVE. 12mo. Cloth. 75

LIFE IN ENGLAND AND AMERICA. Illustrated. 12mo. Cloth. 75

THE VACATION: OR, MRS. STANLEY AND HER CHILDREN. By Mrs. J. Thayer. Illustrated. 18mo. Cloth. Third Edition. 50
THE SAME, Gilt Edges. 75

SUNSHINE AND SHADE: OR, THE DENHAM FAMILY. By Sarah Maria. Fourth Edition. 18mo. Cloth. . 37½
THE SAME, Gilt Edges. 56

THE DREAM FULFILLED: OR, THE TRIALS AND TRIUMPHS OF THE MORELAND FAMILY. 18mo. Cloth. . . 42
THE SAME, Gilt Edges. Fifth Edition. 62½

THE COOPER'S SON: OR, THE PRIZE OF VIRTUE. A Tale of the Revolution. Written for the Young. 18mo. Cloth. Sixth Edition. (In press.) 37½
THE SAME, Gilt Edges. 56

THE SOCIABLE STORY TELLER. Being a Selection of new Anecdotes, humorous Tales, amusing Stories and Witticisms; calculated to entertain and enliven the Social Circle. Third Edition. 18mo. Cloth. 42
THE SAME, Gilt Edges. 62½

TALMUDIC MAXIMS. Translated from the Hebrew; together with other sayings, compiled from various authors. By L. S. D'Israel. 18mo. Cloth. 50
THE SAME, Gilt Edges. 75

LECTURES TO YOUTH. Containing instructions preparatory to their entrance upon the active duties of life. By Rev. R. F. Lawrence. 18mo. Cloth. 50
THE SAME, Gilt Edges. 75

THE SABBATH MADE FOR MAN: OR, INSTITUTED BY DIVINE AUTHORITY. By Rev. Dr. Cornell. 18mo. Cloth. 33⅓
THE SAME, Gilt Edges. 50

CONSUMPTION FORESTALLED AND PREVENTED. By W. M. Cornell, A. M., M. D., member of the Mass. Medical Society. 18mo. Cloth. Fourth Edition. . 37½
THE SAME, Gilt Edges. 56

PASSION AND OTHER TALES. By Mrs. J. Thayer, Author of "Floral Gems," &c. &c. 16mo. Cloth. . . . 62½

TURNOVER. A Tale of New Hampshire. Paper. 25

THE HISTORY OF THE HEN FEVER; A HUMOROUS RECORD. By Geo. P. Burnham. With twenty Illustrations. 12mo. Cloth. 1 25

The work is written in a happy but ludicrous style, and this reliable history of the fowl *mania* in America, will create an immense sensation. — *Courier.*

NEW MINIATURE VOLUMES.

THE ART OF CONVERSING. Written for the instruction of Youth in the polite manners and language of the drawing-room, by a Society of Gentlemen; with an illustrative title. Fourteenth Edition. Gilt Edges. 37½
THE SAME, Gilt Edges and Sides. 50

FLORAL GEMS: OR, THE SONGS OF THE FLOWERS. By Mrs. J. Thayer. Thirteenth Edition, with a beautiful frontispiece. Gilt Edges. 37½
THE SAME, Gilt Edges and Sides. 50

THE AMETHYST: OR, POETICAL GEMS. A Gift Book for all seasons. Illustrated. Gilt Edges. 37½
THE SAME, Gilt Edges and Sides. 40

ZION. With Illustrative Title. By Rev. Mr. Taylor. 42
THE SAME, Gilt Edges and Sides. 56

THE TRIUNE. With Illustrative Title. By Rev. Mr. Taylor. 37½

TRIAD. With Illustrative Title. By Rev. Timothy A. Taylor. 37½

TWO MOTTOES. By Rev. T. A. Taylor. . . . 37½

SOLACE. By Rev. T. A. Taylor. 37½
THE SAME, Gilt Edges and Sides. 50

SONNETS. By Edward Moxon. 31¼
THE SAME, Gilt Edges and Sides. 50

GRAY'S ELEGY, AND OTHER POEMS. The Poetical Works of Thomas Gray. "Poetry — Poetry; — Gray — Gray!" [Daniel Webster, the night before his death, Oct. 24, 1852.] . 31
THE SAME, Gilt Edges and Sides. 50

The following Writing Books are offered on Liberal Terms.

FRENCH'S NEW WRITING BOOK, with a fine engraved copy on each page. Just published, in Four Numbers, on a highly-improved plan.

No. 1 Contains the First Principles, &c. 10
No. 2 A fine Copy Hand. 10
No. 3 A bold Business Hand Writing. 10
No. 4 Beautiful Epistolary Writing for the Lady. 10

James French & Co., No. 78 Washington street, have just published a new series of Writing Books for the use of Schools and Academies. They are arranged upon a new and improved plan, with a copy on each page, and ample instructions for learners. We commend them to the attention of teachers and parents. — *Transcript.*

They commence with those simple forms which the learner needs first to make, and they conduct him, by natural and appropriate steps, to those styles of the art which indicate the chirography not only of the finished penman, but which are adapted to the wants of those who wish to become accomplished accountants. — *Courier.*

A new and original system of Writing Books, which cannot fail to meet with favor. They consist of a series, and at the top of each page is a finely-executed copy. We cordially recommend the work. — *Bee.*

It is easily acquired, practical and beautiful.—*Fitchburg Sentinel.*

We have no hesitation in pronouncing them superior to anything of the kind ever issued. — *Star Spangled Banner.*

FRENCH'S PRACTICAL WRITING BOOK, for the use of Schools and Academies; in Three Numbers, with a copy for each page.

No. 1, Commencing with the First Principles. 10
No. 2, Running-hand copies for Business Purposes. 10
No. 3, Very fine copies, together with German Text and Old English. 10

BOSTON SCHOOL WRITING BOOK, for the use of Public and Private Schools ; in Six Numbers, with copies to assist the Teacher and aid the Learner.

No. 1 Contains the Elementary Principles, together with the Large Text Hand. 10

No. 2 Contains the Principles and First Exercises for a Small Hand. 10

No. 3 Consists of the Capital Letters, and continuation of Small Letters. 10

No. 4 Contains Business-hand Copies, beautifully executed. . 10

No. 5 Consists of a continuation of Business Writing, also an Alphabet of Roman Print. 10

No. 6 Contains many beautiful specimens of Epistolary Writing, also an Alphabet of Old English and German Text. . . . 10

LADIES' WRITING BOOK, for the use of Teachers and Learners, with three engraved copies on each page, and the manner of holding the pen, sitting at the table, &c., explained. 13

GENTLEMEN'S WRITING BOOK, for the use of Teachers and Learners, with three engraved copies on each page, and the manner of holding the pen, sitting at the table, &c., explained. 13

YANKEE PENMAN, Containing 48 pages, with engraved copies. 33

FRENCH'S EAGLE COVER WRITING BOOKS, made of fine blue paper, without copies. 7

www.ingramcontent.com/pod-product-compliance
Lightning Source LLC
LaVergne TN
LVHW020216110826
845151LV00003B/734

* 9 7 8 1 4 2 5 5 3 3 9 5 3 *